AF553265

Chemical Restraint of Wild Animals

NIPA® GENX ELECTRONIC RESOURCES & SOLUTIONS P. LTD.
New Delhi-110 034

About the Authors

Dr. Naveen Kumar presently working as Professor and Head, Veterinary Clinical Complex, Apollo College of Veterinary Medicine, Jaipur, Rajasthan. After more than 34 years of service, he is superannuated as Principal Scientist from ICAR-Indian Veterinary Research Institute, Izatnagar, Uttar Pradesh. Educated at College of Veterinary Sciences, Bikaner Rajasthan, receiving BVSc & AH degree and MVSc degree in Veterinary Surgery and Radiology with First Class. He selected in Agricultural Research Services (ARS) in ICAR and joined at Indian Veterinary Research Institute as Scientist. He is the life member of Association of Indian Zoo and Wildlife and attended several conferences on wild life. He is also associated with the teaching of wild life courses. He is associated with more than 12 extramural funded projects and 10 institute projects. He has been instrumental for procuring 150 lakhs from Department of Biotechnology for development of facilities for Biomaterials and Bioengineering research at IVRI. He has published more than 300 research/clinical papers on Biomaterials, Bioengineering and Veterinary Sciences especially in Veterinary Surgery and Radiology in National and International Journals of repute. He guided 09 PhD and 11 MVSc students and co-advisor of more than 60 students. He has published more than 35 books and manuals and has contributed more than 40 chapters in compendiums and books. He also contributed 8 chapters to books published by international publishers. He received several awards which included Fellow award from National Academy of Veterinary Sciences, Fellow award from Society for Biomaterials and Artificial Organs-India, Gangman Sharma Award, Smt. Ramani Ramchandran Award and Best paper presentation award in various conferences. He has been invited speaker in various international conferences held at Kathmandu (Nepal), Chengdu (China), Hong Kong and Singapore. He was an Expert member in DBT task force for evaluation of new project proposal submitted to department of Biotechnology, Government of India. He is a life member of several professional societies, including the Indian Society for Veterinary Surgery, Indian Society for Veterinary Medicine, Indian Science Congress Association, Indian Carbon Society, Indian Society for Advancement of Canine Practice, Society for Biomaterials & Artificial Organs-India and Society for Tissue Engineering and Regenerative Medicine (India)

Dr. Vineet Kumar has been serving as Associate Professor (Veterinary Surgery and Radiology) at College of Veterinary and Animal Sciences, Bihar Animal Sciences University, Kishanganj, Bihar, since April 18, 2023. Previously, he served as Assistant Professor (Veterinary Surgery and Radiology) at Junagadh Agricultural University, Junagadh, Gujarat, from September 10, 2012 to June 12, 2018, and Sardar Vallabhbhai Patel University of Agriculture and Technology, Meerut, Uttar Pradesh, from June 13, 2018 to April 17, 2023. He completed his BVSc&AH from Bangalore Veterinary College, Karnataka Veterinary, Animal and Fisheries Sciences University in 2008, and his MVSc and PhD (Veterinary Surgery and Radiology) from ICAR-Indian Veterinary Research Institute, Izatnagar, Uttar Pradesh, in 2010 and 2015, respectively. His area of interest is tissue engineering and reconstructive surgery. He has contributed to the development of several decellularization techniques for animal tissues for soft tissue repair. In his service period of over 11 years, he has guided 9 MVSc students, authored 5 books and 34 book chapters, published over 75 research papers in high-impact journals, including 30 papers in international journals (some of them have been extensively cited; citations totalling over 630, H index-15, i10 index-20), written 19 manuals, handled 6 institute-funded research projects, received 9 awards and fellowships, served as a reviewer for several international journals, and been involved in the treatment of various animals, including wildlife animals such as lions, elephants, monkeys, deers, nilgai, and birds.

Dr. Dayamon D. Mathew is currently working as an Assistant Professor, in the Department of Veterinary Surgery and Radiology, Faculty of Veterinary and Animal Sciences, Banaras Hindu University. He obtained his bachelor's degree in Veterinary Science and Animal Husbandry from the College of Veterinary and Animal Sciences, Wayanad, Kerala Agricultural University. He completed his master's degree in Veterinary Surgery and Radiology from Veterinary College, Bangalore, Karnataka Veterinary Animal and Fisheries Sciences University and the research was in canine orthopedics. Later he joined Indian Veterinary Research Institute, Bareilly, Uttar Pradesh for his PhD in Veterinary Surgery and Radiology. His doctoral research was in tissue engineering and regenerative medicine at Dr. Naveen Kumar's Biomaterials and bioengineering laboratory, Division of Surgery, IVRI. After completion of his doctoral degree he joined Sree Chitra Tirunal Institute for Medical Sciences and Technology, Thiruvananthapuram, Kerala as Chitra High Value Fellow for his post-doctoral training. Dr. Mathew then joined the Department of Animal Husbandry, Government of Kerala, as Veterinary Surgeon. He then opted for a deputation to the Kerala Forests and Wildlife Department, Govt. of Kerala and joined as Assistant Forest Veterinary Officer. He was in-charge of wildlife health in two southern most districts of Kerala, Thiruvananthapuram and Kollam. He had Elephant Rehabilitation Center, Kottoor, Thiruvananthapuram, Lion Safari Park, Neyyar, Thiruvananthapuram and Crocodile Park, Neyyar, Thiruvananthapuram under his direct charge. During his tenure in the Kerala Forests and Wildlife department, he earned experience in both captive and free ranging wild animals. Later in 2019, he joined Banaras Hindu University. In Banaras Hindu University, he is a course teacher for different courses of Postgraduate diploma in Wildlife Health Management apart from teaching Veterinary Surgery and Radiology.

Dr. Rahul Kumar Udehiya is an accomplished Assistant Professor in the Department of Veterinary Surgery and Radiology at the Faculty of Veterinary and Animal Sciences, Institute of Agricultural Sciences, Banaras Hindu University. With over 11 years of extensive experience in Veterinary Surgery, Radiology, and Anaesthesia, he has made significant contributions to the field through his dedication to teaching, research, and clinical practice across various prestigious institutions in India. Dr. Udehiya holds postgraduate and doctoral degrees in Veterinary Surgery and Radiology from Guru Angad Dev Veterinary and Animal Science University and the Indian Veterinary Research Institute, respectively. His research focuses on diagnostic imaging, veterinary surgery, anaesthesiology, regenerative medicine, and wild animal surgery. He has contributed to more than 8 chapters in textbooks, published over 60 research articles, and presented more than 50 abstracts at various symposia and conferences, earning several presentation awards. As a recognized expert in his field, he has received the Associate Fellow award from the Indian Society for Veterinary Surgery and Radiology. Dr Udehiya has also mentored and co-mentored nearly 10 students for their masters and doctoral degrees and has successfully managed numerous research projects as both Principal and Co-Principal Investigator. Additionally, he is a member of the PG diploma in wildlife health management running in the Department of Veterinary Surgery and Radiology at the Faculty of Veterinary and Animal Sciences, Institute of Agricultural Sciences, Banaras Hindu University.

Dr. Anil Kumar Gangwar started his career as Assistant Professor, selected as Associate Professor in the year 2012 and joined as Professor in 2015 in the ANDUAT, Kumarganj, Ayodhya. Later he worked as Registrar, DDR and presently serving as Professor and Head, Department of Surgery and Radiology and Director Research in the same university. Dr Gangwar is the Fellow of Indian Society for Veterinary Surgery and Associate Fellow of National Academy of Veterinary Sciences (India). Dr Gangwar is teaching the wildlife courses to the UG students and actively involved in the wildlife surgery. Dr Gangwar developed decellularized scaffolds and used for reconstruction of large hernias in animals first time globally. The work has been published in high Impact factor international journals. He guided 12 MVSc and 03 PhD students and published 103 research papers in reputed peer reviewed journals, 6 books and 21 book chapters. His books are recommended as suggested reading in revised syllabi of post-graduate programmes, ICAR, New Delhi. He is the Editorial board member and reviewer of reputed journal published by Wiley, Elsevier etc. He is the recipient of various prestigious awards including best teacher award.

Chemical Restraint of Wild Animals

Naveen Kumar
Professor and Head
Veterinary Clinical Complex
Apollo College of Veterinary Medicine
Agra Road, Jaipur, Rajasthan

Vineet Kumar
Associate Professor
Department of Veterinary Surgery and Radiology
College of Veterinary and Animal Sciences
Bihar Animal Sciences University, Kishanganj, Bihar

Dayamon D. Mathew
Assistant Professor
Department of Veterinary Surgery and Radiology
Faculty of Veterinary and Animal Sciences
Institute of Agricultural Sciences
Banaras Hindu University, Mirzapur, Uttar Pradesh

Rahul Kumar Udehiya
Assistant Professor
Department of Veterinary Surgery and Radiology
Faculty of Veterinary and Animal Sciences
Institute of Agricultural Sciences
Banaras Hindu University, Mirzapur, Uttar Pradesh

Anil Kumar Gangwar
Professor and Head
Department of Veterinary Surgery and Radiology
College of Veterinary Science and Animal Husbandry
Acharya Narendra Deva University of Agriculture and Technology
Kumarganj, Ayodhya, Uttar Pradesh

NIPA® GENX ELECTRONIC RESOURCES & SOLUTIONS P. LTD.
New Delhi-110 034

NIPA® GENX ELECTRONIC RESOURCES & SOLUTIONS P. LTD.

101,103, Vikas Surya Plaza, CU Block
L.S.C. Market, Pitam Pura, New Delhi-110 034
Ph : +91-11-43860225, Mob.: +91 9717133558, 9540816132
E-mail: newindiapublishingagency@gmail.com
Website: www.nipaersources.com

Print ISBN: 978-93-58875-00-3
ebook ISBN: 978-93-58873-38-2

Composed and Designed by NIPA®.

Preface

Wild animals are more often immobilized for a variety of reasons, including routine health check-ups (veterinary care, surgical intervention, vaccination, hormonal implantation, or sterilization), research purposes (assisted reproductive techniques, blood collection for genetic health analysis, radio-collaring), conservation and management purposes (capture of problematic animals, translocation, and reintroduction), and marking of animals. Thus, the capture and handling of wild animals play a crucial role in wildlife conservation and management. Physical restraint has several limitations and is often unachievable, despite being easy to use and not posing much risk to the animals. Chemical restraint is the most reliable alternative to restraining wild animals and has been employed to relocate or treat animals, resulting in the rescue of many rare and critically endangered wild animals. Chemical restraining of wild animals is a huge responsibility and requires expertise to avoid losses caused by injuries, capture myopathy, and the incorrect use of tranquilizing and immobilizing drugs. Many wildlife veterinarians and forest managers lack such expertise and experience.

The book "***Chemical Restraint of Wild Animals***" offers a comprehensive ready reckoner for wildlife veterinarians and forest rangers on wild animal restraint and very practical information with a systematic documentation of species-specific immobilizing drugs, their administration techniques, monitoring, and potential complications. The book also provides information on a variety of anesthetic equipment, its usage, darting techniques, and precautions while restraining a wide range of wildlife species, including aquatic animals, aves, reptiles, rodents, carnivores, insectivores, ruminants, and large mammals like elephants and rhinoceroses.

It is hoped that this book will be useful for wildlife veterinarians and forest managers for the safe use of chemical restraint in the case of precious wildlife species for their clinical health management, experimentation, and translocation.

Authors

Contents

1

Chemical Restraint of Wild Animals

1.1. Introduction

Wildlife management often requires interventions to safely handle and transport animals. Chemical restraint, also known as immobilization or tranquilization, is a technique widely used to facilitate these processes (Hernandez, 2014). It involves the administration of drugs to temporarily immobilize wild animals, allowing for various procedures such as veterinary medical examinations, tagging, relocation, or rehabilitation. This chapter provides an overview of the principles, methods, drugs, and considerations involved in the chemical restraint of wild animals.

It has been often stated that the most important techniques required by the zoo veterinarian are those used in the capture and restraint of the patient. These techniques have developed into a highly technical science (Clarke *et al.*, 2014). In the early days of zoos, the only practical method was to manually catch and hold the patient. Modern techniques, however, permit a veterinarian to shoot a CO_2 powered gun, which discharges a projectile syringe at an animal resulting in complete immobilization of the animal (Jacques *et al.*, 2009). This is accomplished without any physical contact between the patient and his captors.

1.2. Reasons for Restraint

Reasons for capturing and restraining wild animals vary widely, ranging from routine physical examinations to major surgical procedures. Animals may need to be captured and restrained for purposes such as crating and shipping to another destination. The choice of restraint method depends on factors such as the species of the animal, the specific reason for restraint, and the experience and expertise of the person overseeing the procedure (Kock & Burroughs 2012).

There are numerous methods available for capturing and restraining wild animals, each with its own advantages and risks. Manual capturing, while once common, is now considered highly dangerous for both the animal and the personnel involved (Kock & Burroughs 2012). Therefore, chemical restraint has become the preferred method in many situations due to its safety and effectiveness (La Grange, 2012).

Chemical restraint involves the administration of drugs to temporarily immobilize the animal, allowing for safe handling and procedures without causing undue stress or harm. By immobilizing the animal in a controlled manner, chemical restraint minimizes the risk to both the animal and the personnel involved in the procedure (Kock & Morkel, 2012). This approach is particularly valuable when dealing with potentially dangerous or difficult-to-handle species.

Moreover, various methods for capturing and restraining wild animals, chemical restraint has emerged as a preferred option due to its safety and efficacy. By carefully considering the specific circumstances and needs of each situation, wildlife professionals can ensure the welfare of both the animals and the personnel involved in these procedures.

1.3. Chemical restraint

Wildlife medicine veterinarians must effectively administer anesthetic drugs utilizing various remote delivery systems (Hernandez, 2014). The choice of system depends on the behavior and cooperation of the animal in each situation (Isaza, 2014; Atkinson *et al.*, 2012). Procedures differ based on whether the animals are trained in controlled environments, such as zoos, or free-ranging and unpredictable (Fahlman, 2008).

Chemical restraint is primarily employed for large ungulates and carnivores when physical restraint is not feasible. While it's not suitable for immobilizing large numbers of animals, it becomes the preferred option for individual medical interventions, albeit being expensive and reliant on the veterinarian's expertise (Hernandez, 2014).

Various methods are employed for administering drugs in a chemical restraint procedure, including oral, hand-held injection, pole syringe, and dart usage (Hernandez, 2014; Isaza, 2014). Hand-held injections or pole-syringed administration are preferred for cooperative animals, while remote delivery systems using blow darts or compressed gas projectors are utilized for uncooperative ones (Isaza, 2014).

a) Oral Administration: Oral administration's effectiveness depends on the animal's willingness to accept the drug, making it unreliable due to absorption and effect variability. It's primarily used in carnivores or primates for sedation before darting but necessitates high doses in ruminants due to slow absorption rates (Atkinson *et al.*, 2012).

b) Hand-held Injection: This method is reserved for physically restrained or cooperative animals, requiring the practitioner to administer a rapid intramuscular injection cautiously to prevent self-injection. The drug is drawn into a plastic syringe fitted with a specific needle (Goodman *et al.*, 2013).

c) Pole Syringe: Utilized as an extension of the hand, this method enables drug injection into animals already confined, following similar principles as hand-held injections but with added safety due to a long pole. The syringe, equipped with a lower gauge needle, is attached to the pole and used to inject the drug into the animal's muscle (Hernandez, 2014; Isaza, 2014; Atkinson *et al.*, 2012).

d) Blow Dart (Blow Pipe): Primarily used in controlled environments like zoos, blow darts offer silent projection and reduced impact energy, minimizing potential injuries to animals. They require practiced exhalation for propulsion and are suitable for small and large animals (Hernandez, 2014; Atkinson *et al.*, 2012).

e) Darts: Darts, fired from dart guns, are the predominant method for chemical restraint in free-ranging animals, offering versatility in remote delivery systems. They can be used from various distances and are considered reliable and efficient (Isaza, 2014; Goodman *et al.*, 2013; Atkinson *et al.*, 2012).

Practitioners using dart guns must be experienced in equipment operation, estimating distances, and adjusting charges swiftly to avoid accidents and ensure effective immobilization (Walzer & Gerritsmann, 2015; Hernandez, 2014). Additionally, knowledge of wildlife anatomy, physiology, and behavior is crucial for safe and successful darting procedures (Hernandez, 2014).

Impact energy, determined by dart mass and velocity, influences the effectiveness of dart injection, which is less stressful for animals compared to physical capture methods. Despite being safer and causing low mortality rates, darting is time-consuming, expensive, and requires experienced personnel (Isaza, 2014; Atkinson *et al.*, 2012).

A dart comprises four components: drug storage chamber, discharge method, penetrating needle, and stabilizer for flight balance (Isaza, 2014). Transmitter darts, especially useful in dense vegetation, are supplied by various manufacturers (Kock *et al.*, 2012).

The requirements for the immobilization of wild animal with projectile syringe may be divided into three groups.

A. ***Technical requirements***

1. Highest possible accuracy of aim
2. Greatest possible range

3. Lowest possible energy on impact (Danger of perforation)
4. Simplicity of handling
5. Lowest possible cost

B. Pharmacological requirements

1. Optimal intramuscular action
2. Wide safety of margin
3. Small volume
4. Rapid onset of effect.
5. Effectiveness in as many different genera as possible
6. Simplicity of use

C. Requirements specific to the animal

1. Optimal tolerability
2. Anatomical considerations regarding placing of the shot
3. Choice of suitable time
4. Appropriate preparations before the shooting
5. Choice of place of shooting in relation to absence of disturbance
6. Sufficiently long waiting period after the shooting
7. Optimal familiarity with the animal concerned.

1.4. Types of darts and dart guns

Wildlife chemical immobilization often requires remote drug delivery systems, essential for safely handling and immobilizing animals (Hernandez, 2014). These systems have evolved into highly technical tools, enabling veterinarians to administer drugs to free-ranging animals from a distance, minimizing risks and complications (Walzer & Gerritsmann, 2015; Hernandez, 2014). Remote delivery systems typically consist of a dart and a projector, offering options such as blowpipes, compressed air projectors, or gunpowder cartridge rifles (Hernandez, 2014; Isaza, 2014; Kock *et al.*, 2012).

1. **Pressurized Two-Chambered Compressed Gas Darts:** These darts, made of plastic, feature two chambers separated by a plunger, allowing for the safe delivery of drugs to wild animals (Figure: 1). With careful preparation and testing, veterinarians load the drug solution into the dart, pressurize the posterior chamber, and attach a stabilizer to the back of the dart (Isaza, 2014; Hernandez, 2014). These darts offer lightweight, quiet operation and gentle injection suitable for various species and controlled environments like zoos (Isaza, 2014; Kock *et al.*, 2012b).

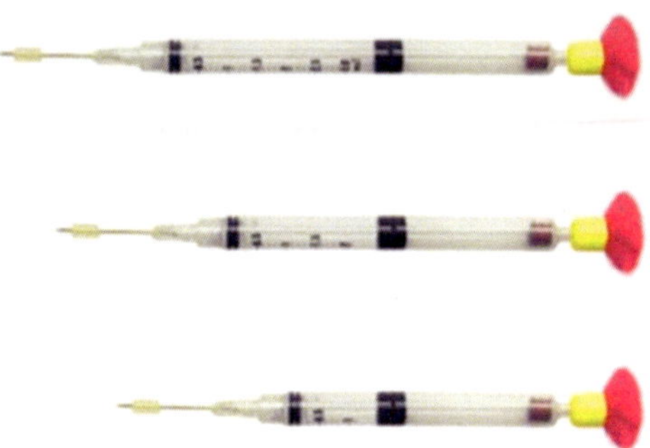

Figure: 1

2. **Gunpowder Explosive Darts:** Utilizing a gunpowder cap mechanism, these darts offer faster injection but pose a higher risk of muscle trauma, especially in small species (Hernandez, 2014; Isaza, 2014; Kock *et al.*, 2012). They come in reusable or disposable models, requiring caution during use to prevent accidental detonation (Figure: 2).

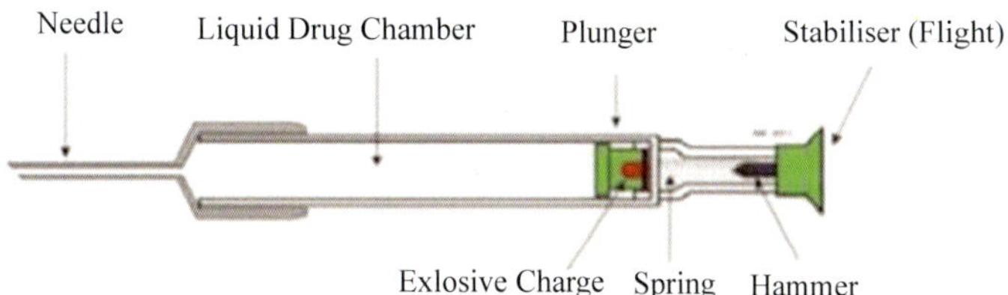

Figure: 2

3. **Remote Drug Delivery Systems:** Blowpipes and gauged blowgun projectors (Figure: 3) are commonly used for short-distance shots in controlled environments like zoos, while air or CO2 rifles and pistols are preferred for free-ranging wildlife (Isaza, 2014). These systems offer highly accurate shots but may cause significant impact and trauma to animals (Isaza, 2014).

Figure: 3

4. **Supplementary Equipment**: Laser range finders, scopes, leather gloves, sights, control pole (often used to control small mammals, reptiles and even medium sized carnivores), baffle boards, squeeze cages and binoculars aid practitioners in precise dart placement, crucial for successful immobilization procedures (Isaza, 2014). Practitioners must be proficient with the equipment, as experience and practice are key to successful immobilization procedures (Kock *et al.*, 2012).

1.5. Needles

Primate needles

Primate needles are specialized 17 gauge needles available in either barbed or plain varieties and come in two lengths: 1/2 inch or 5/8 inches. For large, thick-skinned animals weighing 500 lbs and above, a large diameter of 0.125 inches is typically used. For elephants, the diameter of the needle increases to 0.189 inches.

Additionally, there are various types of needles available with different lengths to suit specific requirements:

NCL-1: 1 ¼ inch long

NCE-1: 1 ½ inch long

NJCL-2: 1 ½ inch long

NCE-2: 2 inches long

NCL-3: 1 ¾ inch long

NCE-3: 2 ½ inches long

These needles offer flexibility in selecting the appropriate length based on the size and species of the animal being immobilized.

1.6. Darting Sites

Identifying appropriate darting sites is crucial to ensure accurate intramuscular (IM) drug administration with minimal risk of injury (Figure 4). Large muscle masses are ideal for darting, with the hindquarters being the preferred target area for most species. However, in some cases where adipose tissue is present, such as at the base of the neck or the triceps muscle, alternative sites may be more suitable (Hernandez, 2014).

1. **Base of the Neck Region:** This area is suitable for animals with heavy necks, such as rhinos, hippos, and large antelopes, as it minimizes muscle trauma. However, there are risks involved, including the possibility of striking cervical vertebrae, vital blood vessels, or the nuchal ligament.

Additionally, darting in this region can be irritating for the animal (Hernandez, 2014; Arnemo *et al.*, 2014).

2. **Triceps Muscle Region:** Considered suitable for most species, the triceps muscle area poses less risk of dart-related complications. However, in thin animals, there's a possibility of the dart hitting the scapula cartilage, scapula spine, or vital blood vessels in the neck. Darting the biceps and triceps is easier from the ground, particularly in large animals like rhinos or elephants. If the animal is too thin, darting the hindquarters is preferable (Hernandez, 2014; Atkinson *et al.*, 2012).
3. **Hindquarters:** Offering ample muscle mass and fewer vital structures, the hindquarters are the most common and desirable darting site. Care must be taken to avoid penetrating the femur and pelvis, especially if darting the medial side of the hind leg. Darting can be performed from the ground or a helicopter, ensuring perpendicular dart placement for deep IM injection. However, practitioners must avoid the soft tissue area of the perineum, particularly in equines (Atkinson *et al.*, 2012).
4. **Withers and Rump:** While providing good muscle mass, these areas can be challenging targets when aiming from the side or above, such as from a helicopter. Nevertheless, they are suitable darting sites, especially in species like large antelopes and rhinos (Atkinson *et al.*, 2012; McTaggart *et al.*, 2012).
5. **Chest:** Darting in the chest region is uncommon due to the risk of hitting the head or penetrating the thoracic cavity, potentially causing pneumothorax. However, in certain situations where no alternative is available, darting in the pectoral area may be considered for species like rhinos, buffalos, giraffes, or large antelopes (Atkinson *et al.*, 2012).

Figure 4: Best darting sites in some of the most common wildlife species

1.7. Common Drugs Used for Immobilizing Wild Animals

1.7.1. Tranquillizers (ataractics and neuroleplics)

Tranquilizers, often mistaken for sedatives, act on adrenergic receptors in the central nervous system (CNS) and peripheral nervous system (PNS), reducing motor activity in animals (Burroughs *et al.*, 2012). Unlike sedatives, tranquilizers selectively suppress behavioral responses and exert potent effects on the autonomic and endocrine systems, with irreversible effects that are not dose-dependent. Increasing tranquilizer dosage prolongs effects, including side effects, without heightening tranquilization levels, contrasting with sedatives. Animals exhibit reduced fear, anxiety, and aggressive behavior, primarily attributed to phenothiazine derivatives like acepromazine (ACP) and butyrophenone derivatives such as haloperidol, commonly used in wildlife practice for immobilization (Lamont & Grimm, 2014). By reducing stress levels and trauma injuries while enhancing adaptation to new environments, tranquilizers significantly improve animal welfare during translocation efforts. However, their action on dopamine receptors and adrenoceptors can lead to adverse effects such as decreased blood pressure, anorexia, convulsions, and thermoregulatory center inhibition, though they may offer anti-emetic properties. Tranquilizers typically lack analgesic effects and antidotes, unlike sedatives, which often do (Lamont & Grimm, 2014; Burroughs *et al.*, 2012).

1.7.2. Phenothiazine derivatives

1.7.2.1. Acepromazine maleate (acetypromazine)

Acepromazine (ACP), a widely used phenothiazine derivative in veterinary practice, is often administered with or without an opioid, historically referred to as neuroleptanalgesia (Lamont & Grimm, 2014; Montané *et al.*, 2003). Phenothiazines typically act by blocking dopamine receptors at the post-synaptic level in the CNS and may inhibit dopamine release (Montané *et al.*, 2003). Common side effects include bradycardia, hypotension, and reduced hematocrit levels. ACP also exhibits antiemetic properties, slows gastrointestinal motility (particularly in equines), induces relaxation, hyperthermia, and protrusion of the third eyelid, with potential extrapyramidal effects like convulsions, circling, and chewing (Lamont & Grimm, 2014). Its effects typically last 6-8 hours; however, caution is warranted in specific immobilization cases, such as semen collection by electro-ejaculation, where ACP may inhibit ejaculation and is therefore not recommended (Plumb, 2008). Structure is identical to promazine except that one hydrogen is substituted by $COCH_3$. Potency is very similar to chlorpromazine, but less toxic. This caused in combination with various drugs induce anesthesia

1.7.2.2. Promazine hydrochioride (sparine)

It has similar properties of chlorpromazine but is less potent, and the absence of a chlorine atom may account for its lack of side effects and toxicity. In contrast to chlorpromazine, it is unlikely to cause hypotension in any species. It is used as preanethetic agent and reduces stress in restrained animals.

1.7.2.3. Chlorpromazine hydrochloride (largactil)

The structure is identical with promazine except presence of chlorine atom by replacing hydrogen atom. Chlorpromazine generally potentiates the action of anesthetics, hyponotics and analgesics and but has wide pharmacological properties. Central nervous system: depresses central nervous system particularly RAS, and hypothalamus and marked antiemetic action, thermoregulatory center depressed.

Cardio vascular system: produce hypotension by depression of vasomotor tone by a central action and peripheral a-adrenoceptor blocking effect. Cardiac output may be increased because of reflex medicated effect on heart.

Respiratory system: depressed or stimulated although drug seems to antagonize respiratory depression associated with sedative analgesics.

Body temperature fall due to hypotension and depression of thermoregulatory centre.

Indications: It is a valuable aid to the anesthetist in providing stress free subject for induction of anesthesia. It reduces the dose of anesthesia. It has also claimed to prevent shock if given prior to anesthesia since a-blockage inhibits sympathetic vascular response.

Contraindications: It should not be used in ruminants as the recovery period is unduly prolonged. If seems to provoke excitement in equines when given intravenously probably due to muscular weakness and hypotension

1.7.3. Butyrophenones derivatives

Each agent of this group is strong anti emetic. They are thought to inhibit the limbic hippocampal system rather than RAS. Effects on cardiovascular and respiratory systems are said to be minimal.

1.7.3.1. Azaperone (suicalm or stresnil)

Substitution of nitrogen atom on the pyridyl ring has produced another butyrophenone. Barbiturates and sedative analgesics are potentiated and they dosage reduced up to 60%. Non toxic short acting and rapidly detoxified and eliminated. Used as excellent preanesthetic especially for pigs.

Azaperone, an agent known to reduce motor activity and inhibit CNS catecholamines dopamine and norepinephrine, causes sedation, vasodilation, and bradycardia, while having minimal effects on respiration, potentially enhancing respiration rate, and possessing antiemetic properties (Plumb, 2008). Despite lacking antagonist or analgesic effects, it may induce chewing, torticollis, catalepsy, and aggressive behavior in certain species like the large antelope (Burroughs *et al.*, 2012; Plumb, 2008). Notably, its vasodilation properties can counteract the vasoconstriction effect of opioids, particularly crucial in species such as the elephant and rhino (Burroughs *et al.*, 2012; Plumb, 2008). Administered via IV, IM, or SQ injections, it exhibits short-acting effects lasting 2-4 hours with a smooth recovery (Wolfe, 2015). Azaperone is commonly combined with opioids like etorphine in immobilization scenarios or used alone as a tranquilizer to enhance welfare during transportation or adaptation to new environments (Lamont & Grimm, 2014).

1.7.3.2. Droperidol (Droleptan) (Insapsine)

It produces mental calmness and indifference. Large doses may cause extra-pyramidal effects. It is the most potent antiemetic so far produced. No effect on respiratory and cardiovascular system. It is 400 times potent than chlorpromazine. It has been extensively used in combination with Fentanyl.

1.7.3.3. Haloperidol

Haloperidol, a butyrophenone derivative, blocks the action of catecholamines like dopamine, inducing sedation. It is primarily utilized independently for the transportation of antelopes as a medium-acting tranquilizer, with an onset of action ranging between 5-10 minutes and lasting up to 8-12 hours (Wolfe, 2015). Unlike some tranquilizers, haloperidol does not cause ataxia or sleepiness. However, in certain antelopes such as nilgai, chinkara, blackbuck, Tibetan antelope, and four-horned antelope, it may provoke aggressive behavior. Notably, haloperidol should not be mixed with opioids, making it unsuitable for use in immobilization cocktails (Hofmeyr, 1981).

1.7.4. Benzodiazepines

The benzodiazepines group is one of the primary sedatives used in veterinary medicine, including wildlife practice. As, zolazepam is combined with tiletamine but other benzodiazepines such as diazepam or midazolam are also available and commonly used in wildlife immobilizations (Burroughs *et al.*, 2012). They promote the action of GABA, which reduces the release or the turnover of acetylcholine in the CNS. It induces depression of the CNS and promotes sedation, skeletal muscle relaxation, anxiolytic and anticonvulsive effects (Plumb, 2008). These drugs have amnesic properties, which are very useful when the veterinarian has to re-immobilize the animal in a short period of time. However, although the properties of sedatives result in amnesic effects in human beings, in felids there is still some reluctance about these effects of benzodiazepines (Ramsay, 2014).

1.7.4.1. Diazepam (valium)

Diazepam, noted as the inaugural sedative employed in early wildlife immobilization procedures (Burroughs *et al.*, 2012), exhibits multifaceted pharmacological properties. Serving as a muscle relaxant, anxiolytic, anticonvulsant, and hypnotic agent, it is also utilized to stimulate appetite, particularly when administered intravenously (Plumb, 2008). However, in certain species, diazepam may induce ataxia, heighten CNS excitement, and prompt behavioral alterations (Plumb, 2008). Due to its potential to precipitate when combined with other drugs, diazepam is unsuitable for inclusion in dart cocktails. Notably, oral diazepam, frequently employed in felids, is administered as premedication with bait, typically 1-3 hours before immobilization (Ramsay, 2014).

1.7.4.2. Midazolam

Midazolam, a benzodiazepine derivative, represents a modern advancement in veterinary pharmacology compared to its predecessor, diazepam (Burroughs *et al.*, 2012). It offers heightened potency and efficacy owing to its more predictable intramuscular absorption profile. This predictability makes midazolam a preferred choice over diazepam in wildlife immobilization practices, where precise and reliable drug delivery is essential (Ramsay, 2014). Its versatility extends to various applications within wildlife management, including serving as a top-up drug during immobilization procedures, acting as a short-acting sedative for rhino relocation initiatives, and even being utilized in baiting strategies (Burroughs *et al.*, 2012). This adaptability underscores midazolam's significance as a valuable tool in the arsenal of wildlife veterinarians and researchers seeking safe and effective methods for animal handling and management.

1.7.4.3. Zolazepam

Zolazepam, an analog of diazepam, shares its anxiolytic properties, making it a valuable asset in veterinary medicine for inducing sedation in animals. This compound is frequently utilized in combination with dissociative anesthetics to achieve balanced anesthesia in exotic species, where precise control over sedation levels is crucial. Additionally, zolazepam's pharmacokinetic profile, characterized by predictable absorption and metabolism, enhances its efficacy and safety in clinical practice. Its use underscores the importance of tailored sedation protocols to ensure optimal outcomes in diverse veterinary settings, particularly when working with exotic or non-traditional animal species.

1.7.5. Benzodiazepine antagonists

1.7.5.1. Flumazenil

Benzodiazepine antagonists, such as flumazenil, play a critical role in countering the sedative effects of benzodiazepines by competing with them at the receptor level within the central nervous system (CNS). This antagonistic action can swiftly reverse the sedation induced by benzodiazepines, whether administered therapeutically or in cases of overdose (Plumb, 2008). Flumazenil, recognized as the primary antidote for benzodiazepine reversal, particularly zolazepam, facilitates a rapid and smooth recovery within 1-2 minutes following intravenous administration (Ramsay, 2014; Plumb, 2008).

1.7.6. Alpha-2 Adrenergic agonists

Alpha-2-agonists and opioids exhibit a synergistic and potentially addictive effect, primarily through their actions at the presynaptic level of noradrenergic neurons. α-2-agonists bind to α-2-adrenoreceptors, inhibiting norepinephrine release and thereby reducing activity in the Sympathetic Nervous System (SNS), leading to decreased heart rate and blood pressure (Lamont & Grimm, 2014). This pharmacological action induces muscle relaxation, sedation, and analgesia, while also mitigating the stress response. However, higher doses of alpha-2-agonists may trigger adverse effects such as vomiting due to chemoreceptor trigger zone activation, hypothermia (typically, though temperature elevation can occur in warm environments), miosis, and hypoxemia. Additionally, inhibition of the antidiuretic hormone may elevate urine production, while decreased gastrointestinal motility could predispose herbivores to bloat and colic issues (Wolfe, 2015; Lamont & Grimm, 2014).

1.7.6.1. *Xylazine hydrochloride (Rompun)*

Xylazine, likely the earliest α-2-agonist introduced in veterinary medicine, enjoys widespread use across species due to its affordability and accessibility (Lamont & Grimm, 2014; Burroughs *et al.*, 2012). It effectively induces muscle relaxation, sedation, and brief analgesia (Burroughs *et al.*, 2012; Plumb, 2008). However, adverse effects such as hyper salivation, muscle tremors (particularly in certain species), gastrointestinal motility suppression leading to ruminal atony and bloat, and vomiting in large cats have been reported. Furthermore, it may induce hypertension followed by hypotension, bradycardia, decreased heart contraction, respiratory depression, and variable body temperature alterations, potentially leading to abortion in late trimesters

(Lamont & Grimm, 2014; Burroughs *et al.*, 2012; Plumb, 2008). Its CNS depressant effects are attributed to decreased norepinephrine and dopamine release and inhibition of norepinephrine release at both presynaptic and postsynaptic adrenoceptors in peripheral vascular smooth muscle (Lamont & Grimm, 2014; Plumb, 2008). Xylazine can be combined with opioids or cyclohexylamines to enhance immobilization efficiency by reducing doses and induction time while promoting improved muscle relaxation (Lamont & Grimm, 2014; Burroughs *et al.*, 2012).

1.7.6.2. *Medetomidine*

Medetomidine, approximately ten times more potent than Xylazine, serves as a sedative with analgesic properties, commonly favored in veterinary settings, particularly among carnivores (Burroughs *et al.*, 2012; Ramsay, 2014). Its selective binding to α-2-receptors necessitates the use of the specific antidote, atipamezole, to reverse its effects. Side effects may include bradycardia, bradypnea, hypothermia, vomiting, and hypotension (Lamont & Grimm, 2014; Plumb, 2008). Typically administered intramuscularly, it is often combined with opioids, ketamine, although its administration can be associated with pain (Lamont & Grimm, 2014; Ramsay, 2014; Plumb, 2008).

1.7.6.3. Detomidine

Detomidine, recognized for its sedative properties alongside its analgesic effects, is commonly employed in veterinary practice (Plumb, 2008). Characterized as a long-acting sedative, it is deemed relatively safe for use in pregnant females and exhibits potency approximately ten times greater than xylazine, akin to medetomidine (Lamont & Grimm, 2014; Burroughs *et al.*, 2012). Although less frequently utilized than medetomidine, detomidine finds application in specific immobilization scenarios, often combined with other drugs for targeted species such as zebra, Indian wild ass and rhinoceros (Lamont & Grimm, 2014; Burroughs *et al.*, 2012).

1.7.7. Alpha-2 Adrenergic antagonists

One of the primary advantages of using α-2-adrenergic agonists lies in their reversible effects facilitated by readily available antidotes (Lamont & Grimm, 2014; Ramsay, 2014; Plumb, 2008). However, veterinarians must consider that the analgesic effects induced by α-2-agonists are also reversed by the antidote, potentially necessitating additional analgesic administration post-reversal. Monitoring for vasodilation and tachycardia resulting from reversal drug use is crucial. Additionally, in certain species, practitioners may opt not to administer the antagonist to facilitate a relaxed recovery and minimize the risk of self-induced trauma lesions (Lamont & Grimm, 2014).

1.7.7.1 Yohimbine

Yohimbine, primarily administered intravenously and gradually, acts as a reversal agent for xylazine-induced sedation but does not counter the effects of medetomidine (Lamont & Grimm, 2014; Plumb, 2008). Notably, its administration may provoke muscle tremors, hyper-salivation, tachypnea, and central nervous system excitement. Additionally, caution must be exercised during its use, as rapid administration can lead to adverse reactions such as transient hypertension, arrhythmias, and even seizures. Furthermore, yohimbine's effectiveness may vary depending on the species and individual response, requiring close monitoring and adjustment of dosage as necessary (Plumb, 2008).

1.7.7.2. Atipamezole

Atipamezole stands as the primary antidote for reversing the effects of medetomidine, attributed to its specific affinity for α-2-adrenergic receptors (Ramsay, 2014; Plumb, 2008). Typically administered via intramuscular (IM) or intravenous (IV) injection, dosage recommendations vary across species, with carnivores requiring 2.5-5 times the milligram equivalent of α-2-agonists, while ruminants necessitate 5 times the equivalent. However, despite its efficacy, atipamezole is relatively costly and may elicit excitement and adverse effects such as vomiting, diarrhea, hyper-salivation, tremors, and aggressive behavior (Plumb, 2008). Notably, in procedures involving pain, additional analgesia like butorphanol is recommended to offset the reversal of the primary analgesic properties of α-2-adrenergic agonists induced by atipamezole (Plumb, 2008; Lamont & Grimm, 2014).

1.7.8. Opioids

The opioid family stands as a cornerstone in the chemical immobilization of ungulates, offering analgesic effects crucial for pain management across various species through distinct opioid receptors (Schumacher, 2008). With mu (μ), delta (δ), and kappa (κ) receptors, each type plays a unique role in mediating analgesia and modifying receptor activity (Lamont & Grimm, 2014; Ramsay, 2008). While μ receptors are chiefly associated with analgesic properties, δ receptors modulate other receptors, and κ receptors provide analgesia in specific CNS and peripheral nervous system areas. Endogenous opioid peptides produced within the organism further contribute to analgesia by coupling with endogenous opioid receptors. Upon systemic administration, opioids exert their effects on CNS receptors, inducing sedation or excitement based on species and dosage. Notably, opioids can impact thermoregulation, causing hypothermia or hyperthermia, stimulate the emetic center, suppress the cough center, alter pupillary diameter, and depress respiratory ventilation, often resulting in bradypnea. Additionally, cardiovascular effects such as bradycardia and arterial hypertension or hypotension may occur, with ruminants predisposed to gastrointestinal complications like bloat or myopathy-related issues (Wolfe, 2015; Lamont & Grimm, 2014; Schumacher, 2008). These challenges are often mitigated by synergistic drug combinations, allowing for lower opioid doses and minimizing adverse effects (Schumacher, 2008).

In veterinary practice, opioids typically consist of morphine derivatives that interact with endorphin receptors in various bodily systems, including the brain, spinal cord, autonomic nervous system, gastrointestinal tract, cardiovascular system, and kidneys, primarily influencing pain perception, behavior, voluntary muscle control, and gastrointestinal motility (Lamont & Grimm, 2014). Their effects range from sedation, excitement, and ataxia to analgesia, respiratory depression, and alterations in cardiovascular and gastrointestinal function. Veterinary opioids are available in agonist, antagonist, or partial agonist/partial antagonist forms, each with a diverse array of effects on the animal (Lamont & Grimm, 2014; Ramsay, 2008).

1.7.8.1. Etorphine hydrochloride (M-99)

In the early 1960s, British scientists synthesized a series of derivatives of the opium alkaloid thebaine, identifying them as potent analgesics and antagonists, with etorphine hydrochloride emerging as the most significant compound. Etorphine acts as a potent central depressant analgesic across most species, binding opiate receptors in various regions of the body and CNS, inducing CNS depression, and sedation while depressing respiratory and cough centers. It exhibits agonistic behavior by interacting with stereo-specific and saturable binding sites in the brain and other tissues, leading to cardiovascular effects like tachycardia and hypertension in ungulates, while causing bradycardia and hypotension in rats, dogs, and monkeys. Additionally, it depresses the respiratory system universally, making it a key component in the immobilization of wild animals either alone or in combination with other drugs. Etorphine's potency enables its use in large species such as elephants and rhinoceros with small volumes easily administered via projectile syringes. However, its addictive properties necessitate careful handling and the availability of antagonists like naloxone, nalarphine hydrochloride, or diprenorphine to counteract accidental injections.

Etorphine is extensively employed in immobilizing ungulates, primarily reversed with diprenorphine hydrochloride. Used mainly for wildlife restraint due to its sedative properties, etorphine's rapid onset leads to ataxia progressing to recumbency, with doses determined by species rather than body mass. To mitigate side effects and improve induction time, it's often combined with tranquilizers or sedatives. Timely administration of reversal agents like diprenorphine, naltrexone, or naloxone is critical to counter etorphine's adverse effects, including respiratory depression, cardiovascular changes, excitement, hypertonicity, hyperthermia, convulsions, and gastrointestinal stasis. In addition to its primary use in wildlife immobilization, etorphine has also shown potential in veterinary anesthesia and analgesia for large animals, particularly in cases where standard sedatives may be ineffective or impractical (Schumacher, 2008; Nielsen, 1999).

1.7.8.2. Thiafentanil oxalate

Thiafentanil oxalate, a synthetic opioid akin to etorphine, shares similar potency and properties with carfentanil and etorphine but boasts a shorter induction time, facilitating swifter immobilization of herbivores with fewer cardiopulmonary depressant effects and a shorter half-life. Compared to etorphine or carfentanil, thiafentanil presents a reduced risk of renarcotization due to its shorter action duration, allowing for effective reversal with antidotes like naltrexone. This makes it a favorable choice for species highly sensitive to etorphine, such as the sable antelope, or in combination with other opioids, as seen in the case of giraffes (Lamont & Grimm, 2014; Janssen *et al.*, 1991).

1.7.8.3. Carfentanil

It is a derivative of fentanyl, it surpasses etorphine in potency, boasting a faster onset of action (approximately 2-5 minutes) but a prolonged duration of action. Unlike etorphine, carfentanil is not entirely reversible with diprenorphine, necessitating naltrexone for efficient reversal, typically in a dosage ratio of 90:1 or 100:1 (Lamont & Grimm, 2014; Caulkett & Arnemo, 2007).

1.7.8.4. Fentanyl citrate (Sublimaze) R33799

Fentanyl citrate, stands as a potent analgesic, surpassing morphine's potency by a factor of 100. Chemically akin to pethidine, it boasts a brief half-life of merely 7 minutes, with a duration of action lasting approximately 30 minutes. Acting on opiate receptors within the CNS, fentanyl induces central depression, often manifesting as ataxia and heightened sensitivity to auditory stimuli across various species. Administration of naloxone and diprenorphine serves to reverse its effects, while adjuncts like azaperone and xylazine mitigate the likelihood of excitation (Lamont & Grimm, 2014; Plumb, 2008).

Upon intravenous administration, fentanyl elicits bradycardia by heightening vagal tone, although its impact on cardiac output and blood pressure remains minor at typical dosage levels. Atropine administration beforehand can preemptively counteract this bradycardic effect, while hypotension may arise with high doses (Lamont & Grimm, 2014; Plumb, 2008). Respiratory function undergoes depression in both rate and depth, particularly when used concomitantly with other respiratory depressants (Lamont & Grimm, 2014; Plumb, 2008).

In veterinary practice, fentanyl is often employed in combination with neuroleptics such as droperidol or fluanisone to achieve neurolept analgesia. One notable formulation, Fentanyl + Droperidol (marketed as Innovarvet) in a 1:50 ratio, comprises 20 mg of droperidol and 0.4 mg of fentanyl per milliliter, with pH adjustment to 3.1±0.4 using lactic acid (Lamont & Grimm, 2014; Plumb, 2008).

1.7.9. Opioid Antagonists

Opioids in wildlife immobilization offer a crucial advantage: their effects can be readily reversed using specific antagonists (Ramsay, 2008; Caulkett & Arnemo, 2007; Nielsen, 1999). These antagonists compete with opioids for receptor sites, with some exhibiting partial agonistic effects, like diprenorphine, butorphanol, nalbuphine, and nalorphine, providing nuanced control over immobilization depth and adverse effects such as respiratory depression (Lamont & Grimm, 2014; Burroughs *et al.*, 2012). Pure opioid antagonists like naltrexone and naloxone offer comprehensive reversal by acting on both μ and κ receptors (Lamont & Grimm, 2014), highlighting the importance of these pharmacological interactions in safe wildlife immobilization practices.

1.7.9.1. Butorphanol tartrate

Butorphanol tartrate, a synthetic partial antagonist, acts as an agonist on κ receptors, providing analgesic effects, while simultaneously functioning as an antagonist on μ receptors, thereby reversing the effects of potent opioids (Lamont & Grimm, 2014; Burroughs *et al.*, 2012). Often administered alongside other drugs like azaperone or medetomidine, its versatility allows for use in standing sedations, facilitating the transportation of rhinos, and enhancing respiration during deep immobilization scenarios involving opioids by modulating anesthesia levels (Miller & Buss, 2015).

1.7.9.2. Diprenorphine

Diprenorphine, a commonly utilized partial antagonist, serves as the primary antidote for etorphine, although it is less effective for reversing thiafentanil. When administered at lower doses, it exhibits antagonistic effects, such as improving respiration following intravenous administration. However, at higher doses (typically at a ratio of diprenorphine 10:1), or when repeated, it can exert agonistic effects. Dosage calculations are based on the specific opioid used for immobilization, typically ranging from 3 times the dose of etorphine for large animals to 2 times the dose for other ungulates, and can be administered via intravenous or intramuscular injection (Lamont & Grimm, 2014; Burroughs *et al.*, 2012).

1.7.9.3. Nalbuphine

Nalbuphine share similar characteristics to butorphanol and are often considered as analogous options (Lamont & Grimm, 2014). Veterinarians utilize these drugs to enhance respiration during deep anesthesia and for procedures involving walking rhinos (Burroughs *et al.*, 2012).

1.7.9.4. Naltrexone

Naltrexone functions as an antagonist against opioid action on μ receptors, distinguishing itself from other mentioned antagonists by its pure nature and extended half-life, which reduces the risk of renarcotization. Effective even in human opioid overdoses, it counters the effects of various opioids, including renarcotization with carfentanil, thereby enhancing the animal's responsiveness, pain perception, and awareness (Lamont & Grimm, 2014). Following administration, re-immobilization is not possible for the next 24 hours (Caulkett & Arnemo, 2007). Typically delivered via IM or IV injections, the recommended ratio is 10:1, although some sources suggest variations depending on the specific opioid (Nielsen, 1999). In the case of rhinos, naltrexone is essential upon release into the field.

1.7.9.5. Naloxone

Naloxone, also a pure antagonist, has historically been the primary choice for addressing human opioid intoxications, although naltrexone can also serve this purpose . Due to its short duration of action, however, there is a risk that the immobilizing opioid may still exert its effects on the animal even after naloxone administration, necessitating careful monitoring to prevent renarcotization (Caulkett & Shury, 2014; Morkel & Kock, 2012). Naloxone effectively reverses the effects of all opioids, enhancing the animal's responsiveness, awareness, perception, and pain sensitivity.

1.7.9.6. Nalorphine

Nalorphine hydrochloride, commonly known as Nalline, is derived from morphine and functions as a partial agonist. It counteracts many of the effects induced by morphine and its derivatives by binding to their receptors. While it can be administered via subcutaneous, intravenous, or intramuscular routes, the intravenous route is preferred for rapid onset of action. Nalorphine typically acts as a narcotic antagonist, but in the absence of other antagonists, it may act as a narcotic itself and potentially exacerbate central nervous system depression. Administration of very large doses can lead to respiratory paralysis. One of nalorphine's primary actions is preventing or alleviating the respiratory depression caused by morphine derivatives. For reversing the effects of etorphine, a ratio of 10-20 mg of nalorphine to 1 mg of etorphine is recommended, administered intravenously.

Caution must be exercised to ensure the proper dosage of nalorphine. If the initial dose fails to alleviate respiratory depression, additional doses should not be administered. In cases of nalorphine overdose, respiratory support measures are essential, including maintaining a patent airway and providing oxygen supplementation.

1.7.10. Cyclohexylamines

The cyclohexylamine group, utilized as dissociative anesthetics in wildlife practice, lacks antidotes for reversal, leading to prolonged recovery times. They are often combined with tranquilizers or reversible sedatives to mitigate side effects such as skeletal muscle hypertonia (Thurmon & Short, 2007). These agents induce catalepsy, analgesia, immobility, loss of consciousness, and amnesia. Despite immobilization, animals retain reflexes and may experience mydriasis or nystagmus (Burroughs *et al.*, 2012; Thurmon & Short, 2007). They are effective for immobilizing carnivores, primates, reptiles, or birds without severe respiratory or cardiovascular effects, even at high doses. Ketamine and tiletamine are common types administered via IM or IV injection, with induction taking 5-10 minutes. Side effects may include convulsions, hyper-salivation, and hyperthermia, necessitating caution and muscle relaxants to reduce risks (Burroughs *et al.*, 2012; Caulkett & Arnemo, 2007). While their effects on excitement are unpredictable, animals typically do not exhibit running behavior seen with opioids in ungulates. Co-administration with α-2-agonists may induce vomiting in some cases (Radcliffe & Morkel, 2014).

1.7.10.1. Ketamine HCL (vetalar, ketalar)

Ketamine, widely used across species including wildlife, is typically administered via IM injection, orally, or IV injection in wildlife practice, inducing complete immobilization within 5-10 minutes. It acts by inhibiting GABA, suppressing serotonin, norepinephrine, and dopamine in the CNS, and activating the limbic system (Plumb, 2008). Dose-dependent effects last 2-3 hours and may include convulsions and hyperthermia, mitigated by combining ketamine with other drugs (Burroughs *et al.*, 2012; Caulkett & Arnemo, 2007). Ketamine induces anesthesia, catalepsy, and amnesia but elevates heart rate, blood pressure, and suppresses respiratory rate (Lamont & Grimm, 2014; Plumb, 2008). While primarily used in carnivores, it can be administered to ruminants intravenously as a top-up drug (Caulkett & Arnemo, 2007; Nielsen, 1999). With a safety margin up to 10 time to the recommended dose, ketamine may induce hallucinations in humans, potentially manifesting as abnormal behaviors and vocalization in primates and felids during recovery (Lamont & Grimm, 2014; Fowler, 2008).

1.7.10.2. Tiletamine HCL (CI-634)

Chemically related to ketamine and has very similar spectrum of activity. Rapid onset and lost for 30-50 min. Used in combination with zolazepam called CI-744. Alone produces bradycardia but combination produces tachycardia with slight increase in blood pressure. Tiletamine, often combined with zolazepam, provides injectable anesthesia with a short induction time of 5-8 minutes, mainly used in carnivores via IM injection (Lamont & Grimm, 2014; Burroughs *et al.*, 2012). Its potency, 3-4x higher than ketamine, offers immobilization lasting 2-4 hours or longer (Plumb, 2008; Walzer & Huber, 2002). Although it may induce muscular hypertonicity, this is rare due to its combination with zolazepam, which reduces convulsions, induces muscle relaxation, and aids recovery (Lamont & Grimm, 2014; Burroughs *et al.*, 2012). Nonetheless, side effects such as hyperthermia, cyanosis, vomiting, and abnormal vocalization may occur (Nielsen, 1999). Long-lasting CNS effects lasting 24-48 hours post-administration can include convulsions, weakness, and anorexia (Fowler, 2008). Often used with an α-2-adrenergic-agonist like medetomidine in felids, allowing for reduced dissociative dose and fewer side effects (Ramsay, 2014).

1.7.11. Other drugs used in wildlife practice

1.7.11.1. Doxapram

Doxapram acts as a temporary CNS stimulant, enhancing respiration by stimulating the medullary respiratory center and peripheral chemoreceptors (Burroughs *et al.*, 2012; Plumb, 2008). Its effects, though rapid, are short-lived following IM or IV administration, necessitating replacement with drugs like butorphanol or naltrexone to maintain respiratory function in hypoxic animals. Caution is advised when using it in rhinos due to potential CNS excitement and increased muscle tremors (Radcliffe & Morkel, 2014).

1.7.11.2. Hyaluronidase

Hyaluronidase, an enzyme that breaks down hyaluronic acid, reduces connective tissue viscosity, and enhances drug diffusion, is utilized in wildlife practice to optimize chemical immobilization of free-ranging animals (Burroughs *et al.*, 2012; Cattet & Obbard, 2010). Incorporated into dart formulations, it accelerates drug absorption in muscles, thus shortening induction times, particularly beneficial in species like nilgai, chinkara, blackbuck, Tibetan antelope, and four-horned antelope (Cattet & Obbard, 2010).

1.8. Complications in wildlife immobilizations

1.8.1. Stress

Stressful procedures like chemical immobilization, physical restraint, or human presence trigger acute stress responses in wild animals, impacting their homeostasis (Arnemo *et al.*, 2014). The hypothalamic-pituitary-adrenal (HPA) axis and the sympathetic nervous system (SNS) are key mediators of stress responses, with the HPA axis orchestrating long-term reactions and the SNS facilitating short-term responses by releasing adrenaline (Hernandez, 2014; Meltzer & Kock, 2012). Elevated glucocorticoid levels from HPA activation have detrimental effects, including reduced fertility, stunted growth, and immune suppression. Adrenaline, released during short-term stress, triggers the "fight or flight" response, increasing heart rate and blood pressure (Meltzer & Kock, 2012). Prolonged stress can lead to complications such as capture myopathy, hyperthermia, and even death (Arnemo *et al.*, 2014).

Stress manifests in three phases: alarm, resistance or adaptation, and exhaustion, with each stage influencing glucocorticoid levels and immune function (Hofmeyr *et al.*, 2012; Meltzer & Kock, 2012). Maladaptation, common in relocated animals, stems from stress and environmental mismatches, often leading to health issues like heartwater in springboks moved to unsuitable habitats (Meltzer & Kock, 2012).

1.8.2. Capture myopathy

Capture Myopathy (CM), also known as Stress Myopathy or Exertional Rhabdomyolysis, is a common metabolic disorder found in both domesticated and wild animals, often resulting in death (Wolfe, 2015; Arnemo *et al.*, 2014). It arises from extended pursuits, restraint techniques, relocation, and other stress-inducing factors (Wolfe, 2015). Intensive muscular activity, triggered by stressors like helicopter chases, leads to exhaustion, reduced oxygen delivery, and increased lactic acid production, resulting in muscle necrosis and myoglobin release into the bloodstream (Wolfe, 2015).

Signs of CM include prostration, ataxia, hyperthermia, tachypnea, muscle stiffness, paralysis, and acute renal failure. Death can occur within minutes to weeks after immobilization (Blumstein *et al.*, 2015; Paterson, 2014). Prevention involves minimizing stressors during immobilization, including reducing pursuit time, temperature, and external stimuli (Arnemo *et al.*, 2014; Hernandez, 2014; Sanchez, 2011).

Treatment focuses on cooling the animal, aggressive fluid therapy, correction of metabolic acidosis, analgesics, muscle relaxants, and vitamins (Wolfe, 2015; Hernandez, 2014; Paterson, 2014). Dantrolene sodium and corticosteroids may also be administered to protect vascular integrity (Paterson, 2014). Despite treatment, success rates are low, emphasizing the importance of prevention (Wolfe, 2015; Paterson, 2014).

CM can be classified into four categories based on its severity and presentation (Blumstein *et al.*, 2015; Wolfe, 2015; Hernandez, 2014).

1. Hyperacute or Capture Shock Syndrome: refers to a terminal stage of organ failure characterized by a drop in blood pressure due to factors such as blood volume loss, severe dehydration, or high parasitic load. It can occur within 1-6 hours post-capture and manifests as tachypnea, tachycardia, hyperthermia, hypotension, depression, and ultimately death (Wolfe, 2015; Hernandez, 2014). Serum enzyme analysis typically reveals elevated levels of aspartate aminotransferase (AST), creatine kinase (CK), and lactate dehydrogenase (LDH), while postmortem findings include hepatic and pulmonary congestion, intestinal congestion, and histological evidence of necrosis in various organs (Wolfe, 2015; Hernandez, 2014).
2. Ataxic Myoglobinuric Syndrome: the most commonly observed CM, may occur hours to days after immobilization. Clinical signs include ataxia, torticollis, and myoglobinuria, with elevated serum enzyme levels and histological evidence of skeletal muscle necrosis and kidney damage (Wolfe, 2015; Hernandez, 2014).
3. Ruptured Muscle Syndrome: typically presents within 24-48 hours post-capture, with clinical signs including hindquarter drooping and hyperflexion of the hock due to muscle rupture (Wolfe, 2015; Hernandez, 2014). Serum enzyme levels are elevated, and lesions consistent with muscle necrosis are found in various muscles (Paterson, 2014).
4. Delayed-Peracute Syndrome: is a rare condition that can occur more than 24 hours after immobilization, resulting in sudden death due to ventricular fibrillation. Elevated serum enzyme levels and mild to moderate rhabdomyolysis may be observed (Wolfe, 2015; Hernandez, 2014).

Differential diagnoses for CM in wildlife include plant toxicity, malignant hyperthermia, early tetanus, hypocalcemia, or myositis (Paterson, 2014).

1.8.3. Renarcotization

Renarcotization is a complication often observed in wildlife medicine, especially in herbivores immobilized with opioids (Napier & Armstrong, 2014; Pas, 2014). After an animal has been immobilized and reversed with an antidote, it may exhibit signs of re-sedation hours later. Antagonists like diprenorphine, used to reverse carfentanil or thiafentanil, can sometimes cause renarcotization. Due to its longer half-life, naltrexone is considered a safer option for re-immobilization without the risk of re-sedation after antidote administration. However, precise dosing of the antidote is crucial to avoid underdosing, which may lead to renarcotization (Napier & Armstrong, 2014).

Opioid levels in the bloodstream can maintain clinical signs for 12-24 hours post-reversal. Symptoms may include excitement, circling, high stepping, and recumbency paddling, potentially resulting in hyperthermia, acidosis, and exhaustion. In such cases, the veterinarian may need to administer half the original antidote dose via intramuscular injection. If the clinical signs persist, subsequent doses may be required until renarcotization ceases (Napier & Armstrong, 2014).

1.8.4. Physical trauma

Physical trauma is a potential consequence of immobilization procedures, whether through physical or chemical means. Animals can sustain injuries such as lacerations, fractures, or abrasions, either accidentally inflicted by themselves, other animals, or the team conducting the procedure. Adequate handling techniques, understanding of animal behavior, and awareness of environmental hazards can help prevent many instances of physical trauma (Arnemo *et al.*, 2014). While some level of loss is expected during immobilization procedures, with approximately 10% of animals experiencing mortality, the loss of even a single individual, particularly in the case of endangered species, can have significant consequences (Hernandez, 2014; Meltzer & Kock, 2012).

1.8.5. Hipoxemia and hypoxia

Hypoxemia, characterized by low arterial oxygen tension, is a common complication during immobilization procedures. It can stem from respiratory depression induced by immobilization drugs or from loss of lung volume due to improper positioning, particularly in herbivores where gastrointestinal content can compress the diaphragm. Pneumothorax, resulting from dart penetration or horn puncture, is another potential cause (Arnemo *et al.*, 2014; Lamont & Grimm, 2014; Napier & Armstrong, 2014). Hypoxia, inadequate oxygen levels in body tissues, can lead to cellular degeneration in vital organs such as the brain, heart, kidneys, and liver. Reduced tissue perfusion or impaired hemoglobin function can exacerbate hypoxia (Fahlman, 2014; Cracknell, 2014).

1.8.6. Hypertermia/hypothermia

Hyperthermia, or overheating, is a common complication during wildlife immobilization, often triggered by high environmental temperatures, exertion, or drugs affecting thermoregulation. Symptoms include panting, weakness, arrhythmia, shallow breathing, and convulsions, with temperatures exceeding 43°C considered critical (Arnemo *et al.*, 2014; Schumacher, 2008). Prompt cooling measures such as shade, cold water, or oxygen administration are vital to prevent fatalities (Hofmeyr *et al.*, 2012). Hypothermia, with temperatures below 35°C, is less frequent but can occur in cold, wet environments, particularly affecting young or debilitated animals. Warmth and shelter are essential for recovery and prevention (Arnemo *et al.*, 2014; Schumacher, 2008).

1.8.7. Bloat

In ruminants, maintaining a correct sternal recumbency position during immobilization is crucial to prevent bloat, which can be induced by drugs like α-2-agonists causing ruminal atony. Bloat results from gas accumulation due to the inability to eructate. If bloat occurs, repositioning the animal sternal with the head down helps drain saliva. Intubation or rumen trocarization may be necessary, with antidotes for α-2-agonists used if applicable (Arnemo *et al.*, 2014; Napier & Armstrong, 2014; Hofmeyr *et al.*, 2012).

1.8.8. Vomit/regurgitation and aspiration pneumonia

Vomiting and regurgitation are treated as emergencies due to the risk of aspiration pneumonia. These can occur as side effects of opioids or α-2-adrenergic-agonists. Correct recumbency positions differ between species, with ruminants positioned sternal and carnivores lateral. If aspiration of stomach/rumen content occurs, leading to pneumonia risk, immediate administration of a broad-spectrum, long-acting antibiotic is necessary (Arnemo *et al.*, 2014; Pas, 2014; Hofmeyr *et al.*, 2012).

1.8.9. Equipment failure

Equipment failure is a significant concern during wildlife immobilization procedures. Despite careful planning and drug selection, complications can arise due to various factors, including inexperienced personnel. Dart equipment, essential for such procedures, can encounter several issues during use. Needles may bend or break upon impact, leading to inadequate drug administration. Improper loading of darts can result in incorrect projection, while loss of tailpieces can disrupt balance. Weather conditions may affect power charges, and gas cylinders may leak. Missing the target or hitting the wrong area can also occur, potentially causing complications. Factors such as impact strength and velocity can influence drug injection. Rusty or dirty equipment, along with dart wounds, may lead to infections, while misplaced darts could cause pneumothorax (Arnemo *et al.*, 2014; Hernandez, 2014; Atkinson *et al.*, 2012).

2

Chemical Restraint of Aquatic Animals

2.1. Classification of fish

Fish are incredibly diverse aquatic vertebrates. There are over 32,000 described species of bony fish, over 1,100 species of cartilaginous fish, and over 100 hagfish and lampreys. Fish can be broadly classified into three groups:

2.1.1. Superclass Agnatha (Jawless fish): These are the most primitive fish, lacking jaws and paired fins. They have existed for over 500 million years. Examples include lampreys and hagfish.

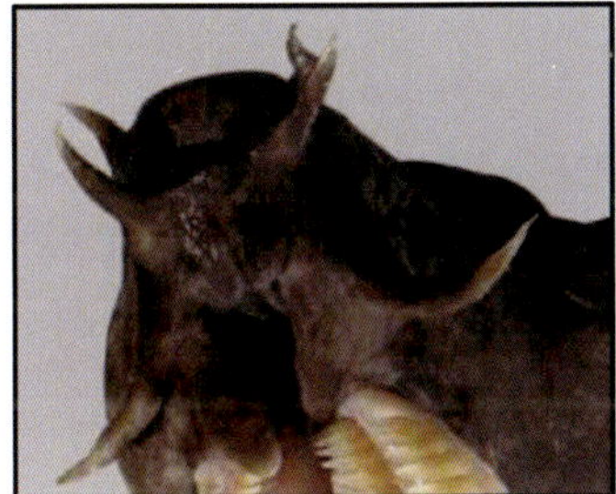

Hagfish

2.1.2. Class Chondrichthyes (Cartilaginous fish): These fish have skeletons made of cartilage instead of bone. They have jaws and paired fins. Examples include sharks, skates, and rays.

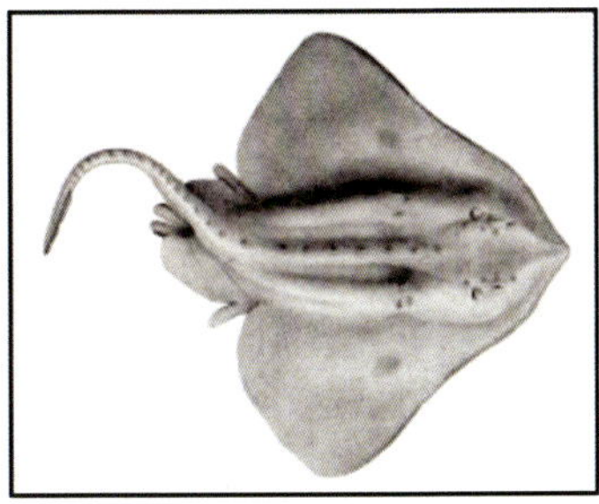

Skate fish

2.1.3. Superclass Osteichthyes (Bony fish): This is the largest and most diverse group of fish. They have skeletons made of bone and have jaws and paired fins. Examples include tuna, salmon, goldfish, and many more.

Salmon fish

Fishes should be handled only when necessary and then with a soft net. Hand catching is strictly to be avoided, in as much as fish are often injured in the process. Physical restraint of fish without the aid of drugs is always attended by some degree or risk.

They are water breathing, poikilotherms dependent on gas exchange across the gill lamellae and there for liable to rapid suffocation when removed from their proper environment. Trauma induced by intense muscular activity and loss of the protective mucilaginous covering renders them susceptible to bacterial and fungal invasion.

- Short periods of immobilization may be required for taking measurements, fin clipping, tagging and local treatment of skin lesions.
- Sedation for many hours may be required for live transportation without fatal stress.
- Relatively deep anesthesia may be required for experimental surgery and other surgical procedures.

Early work in including anesthesia in fish was done by traumatic anesthesia, electric anesthesia or by adding brandy to the aquarium water. Later more methods of anesthetizing fish were developed by using inhalant anesthetic agents. Fish anesthesia is a crucial process used for various purposes in fish health management. Here's a breakdown of the different aspects of fish anaesthesia.

2.2. Reasons for fish anaesthesia

- ***Pain relief during surgical procedures:*** Fish can feel pain, and anaesthesia is essential to ensure they don't suffer during surgeries like tumour removal or treatment of internal injuries.

- ***Immobilization for examinations or minor procedures:*** Anaesthesia helps keep fish still during examinations or minor procedures like fin clipping or scale scraping, making the process less stressful for both the fish and the person handling it.
- ***Transport sedation:*** For short-distance transport, mild sedation can minimize fish stress.
- ***Euthanasia:*** An overdose of certain anaesthetics agents can be used for humane euthanasia of fish.

2.3. Types of fish anaesthetics

There are two main delivery methods for fish anaesthesia:

1. ***Immersion anaesthesia:*** The most common method, where fish are placed in a separate tank containing a measured concentration of anaesthetic.
2. ***Inhalation anaesthesia:*** Less commonly used, but can be useful for certain procedures. It involves anesthetizing the fish with a gas anaesthetic vaporized into the water.

The most commonly used fish anaesthetics include:

- ***Tricaine methane sulfonate (MS-222):*** (3-aminobenzoic acid ethyl ester methane sulfonate) A widely available and versatile anaesthetic used for immersion anaesthesia. The induction of anaesthesia is rapid. It is considered to be safe for the handlers but can irritate while coming in contact with eyes and mucous membranes. Generally, tricaine is used at a dose rate of 25-100 mg/L and the dose is related to species, size and density of the fish and the temperature of the water. Excessive exposure for a longer period can cause mortalities, therefore it should be avoided.
- ***Benzocaine:*** (p-aminobenzoic acid ethyl ester): Dose rate is 25-100 mg/L. Usually it may take around 4 minutes or lesser for induction of anaesthesia. Benzocaine is considered to be harmless to handlers but can cause respiratory irritation.
- ***Lidocaine:*** (2-diethylamino-N-(2,6-dimethylphenyl) acetamide). Lidocaine is usually combined with sodium bicarbonate (1g/L) for anaesthesia.
- ***Metomidate and Etomidate:*** (1-(1-phenylethyl)-1H-imidazole-5-carboxylic acid methyl and ethyl ester respectively). The dose is at a rate of 1-10mg/L. Induction can be achieved within 3 minutes while the recovery is prolonged. This can cause twitching of the muscle as a side effect while metomidate is used, while it is not reported with etomidate.

- ***Propoxate:*** (Propyl-DL-1-(phenylethyl) imidazole-5-carboxylate hydrochloride). It is 10x time more potent that tricaine. The dose is 0.5 mg/L to 10 mg/L. At higher doses, this can cause respiratory arrest.
- ***Ketamine hydrochloride:*** (2-(0-chlorophenyl)-2-(methyl-amino) cyclo-hexanone hydrochloride). The margin of safety for ketamine is wide. This is administered intravascularly at a dose rate of 30 mg/kg usually, it takes around 3 minutes for induction and 1-2 hours for recovery. This can cause some degree of stress during the early stages of anaesthesia but may not cause a ventilatory block. Ketamine is a drug of choice for long-term anaesthesia.
- ***Quinaldine sulphate:*** (2-methylquinoline sulphate). This is used for short-term anesthesia. This is not used for surgical manipulation as it doesn't block the involuntary muscular movement completely.
- ***Propanidid:*** (4-(2-(diethylamino)-2-oxoethoxy)3-methoxybenzene-acetic acid propyl ester). This can be given intraperitoneally at a dose of 2 mg/kg or at a dose of 1.5-3 ml/L. The induction happens within 4 minutes while recovery in 5-10 minutes.
- ***Eugenol:*** A natural anaesthetic derived from clove oil, often used for short procedures.
- ***Isolectin:*** An anaesthetic extracted from salmon skin, with growing interest due to its safety profile.

2.4. Factors affecting fish anaesthesia

- ***Fish species:*** Different fish species have varying sensitivities to anaesthetics.
- ***Water quality:*** Water parameters like temperature, pH, and oxygen levels can affect how fish metabolize anaesthetics.
- ***Fish health:*** Fish that are already stressed or ill may be more susceptible to the effects of anaesthesia.

2.5. Safe handling of fish anaesthetics

Fish anaesthetics can be harmful if not handled properly. It's important to:

- Use only under the guidance of a veterinarian or trained aquarist.
- Follow recommended dosages carefully.
- Monitor fish closely during anaesthesia and recovery.
- Dispose of waste anaesthetic solution responsibly.

By following these guidelines, fish anaesthesia can be a safe and effective way to ensure the well-being of fish during various procedures.

2.6. General considerations during anaesthesia in fishes

As in other animals, fasting may be done in fish as well prior to administration of anaesthesia. Skipping a feeding cycle prior to anaesthesia may be followed. Sometimes, a fish that is fed, may regurgitate and may lead to fouling of the water. For this reason, it is always better to have different water tanks or supplies for anaesthesia induction, maintenance and recovery. Reducing the stress while handling the fish can be beneficial for a smooth induction and recovery from anaesthesia. Too much physical restraint of fish can cause abrasions and cause the mucus over the body to be lost. A well-aerated anaesthetic water can help in reduced breathing. If the fish is taken out of water for examination/ procedure, moistening the skin and fins should be done.

Iwama et al (1989) described about the stages of anaesthesia and recovery in fish. Accordingly, stages of anaesthesia can be classified into three viz., I, II and III. Stage I is characterized by loss of equilibrium, Stage II by loss of gross body movements but with continued opercular movements and Stage III with no opercular movements. Stages of recovery in fish are classified as Stage I, II and III where the body is immobilized but opercular movements just starting, regular opercular movements and gross body movements beginning and equilibrium is regained and preanesthetic appearance respectively.

2.7. Recommended anesthetic techniques

Fishes are sensitive to abrupt change in water temperature, pH, and dissolved minerals. It is recommended that some of the resident tank water should be included in the anesthetic solution. Containers should be constructed of chemically inert substances such as polyethylene and metal fixtures, partially copper and zinc should be avoided.

Fish should be fasted for 24-48 hours since they otherwise vomit and gill lamellae may become blocked by solids. Small species can be netted and transferred to the anesthetic container while large species can be netted and anesthetized by immersion of head and gills, or alternatively by spraying the gills with anesthetic solution.

After an initial phase of excitement lasting only a few seconds followed by rather erratic swimming, a fairly standard pattern of narcosis is observed, when fish are allowed to swim freely in an anesthetic solution. The behavioral changes associated with six levels of narcosis have been described by McFarlane and Klontz (1969).

Behavioral changes at various levels of narcosis

(McFarlane and Klontz, 1969; taken from "Animal Anesthesia" book by C.J. Green, 1982)

Levels	behavioral changes
Normal	Relative to external stimuli, muscle tone is normal
Light sedation	Equilibrium normal, slight loss of reactivity to external, visual tactile stimuli
Deep sedation	Equilibrium normal, total loss of relatively to external stimuli except strong pressure, slight decrease in opercular rate
Partial loss of equilibrium	Partial loss of muscle tone, swimming erratic, increases in opercular rate, reactive only to strong tactile and vibration stimuli
Total loss of equilibrium	Total loss of muscle tone and equilibrium, rapid opercular rate, reactive only to deep pressure stimuli
Loss of reflex reactivity	Total loss of reactivity, opercular movements very shallow, heart rate very slow
Medullar collapse	Opercular movement cease immediately after gasping followed by cardiac arrest

The initial agitation is followed by gradual loss of equilibrium, swimming movements become less frequent, and the fish turn upside down and then usually swims to the bottom to rest on its back. Total time to reach this stage depends upon agent and concentration. Ex. Tricaine methane sulfonate (MS-222), dose is 100 mg/L, it takes 60-90 seconds. Level 4 is adequate for most surgical procedures. At this level, the fish can be removed from the solution for maintenance during manipulating procedures (At this stage-complete muscle relaxation and diminished but not necessarily abolished reaction if the tail base is squeezed).

If muscle spasms are seen during the procedure, it can simply be returned to the anesthetic solution for a few seconds to deepen narcosis. A relatively steady level of narcosis can be maintained for prolonged periods provided the fish is kept moist by wrapping it in damp cloths or laying in a V-shaped trough constructed from sponges in a shallow tray. The head and gills can be totally immersed in anesthetic solution or they may be intermittently sprayed or continuously irrigated by gravity drip feed over the buccal surface. If surgical anesthesia is required for many hours, it is necessary to control water temperature, provide aeration and remove waste products. Continuous irrigation with fresh anesthesia solution draining from the tray is a simple way to satisfy these requirements. Alternatively, the solution may be recycled after filtration and oxygenation in a continuous perfusion circuit. Overexposure to anesthetic and imminent medullary collapse are signaled by cessation of regular opercular movement, with the occasional exaggerated respiration, and flaring of the opercular eventually becoming weaken until they cease altogether. Cardiac arrest is then likely to occur within 60-90 seconds unless emergency action is taken. The anesthetic can be rapidly reversed by irrigating the buccal cavity with fresh water, immersing the fish and moving it backward and forwards with its mouth open. Water must not be pumped through the opercular opening to flow over the gills in an anterior direction and hence out of the mouth, since very little oxygen will be taken up into the lamellar circulation and the fish will rapidly die from hypoxia. Recovery is complete within 5-30 min after fish are removed from the anesthetic solution and returned to clean water (Jolly *et al.,* 1972). Recommend washing the fish in an intermediate container before returning it to the resident tank to avoid contamination of the latter and possibly delaying recovery. Food should be withheld for 24-48 hours after prolonged anesthesia, but this is not necessary after short period of narcosis.

Following methods are recommended in order of preference:

1. Fresh solution of 200 mg Benzocaine in 5 ml acetone and use it on the same day. This is added to 8 liters of water, the benzocaine concentration 25 ppm. Thoroughly stirred before use.
    - 20-30 ppm- provides sufficient sedation for tagging, marking and measuring fish.
    - 50 ppm- required for surgical anesthesia, induction time is 1-2 minutes, recovery is 3 min after removal from the solution, probably because of its neutral pH benzocaine does not produce the initial excitement seen when fish are immersed in unbuffered tricaine. The cost of benzocaine is very much less than tricaine
2. Fresh solution of tricaine prepared immediately before use. 25-300 mg/L of water depending upon the required level of narcosis.
    - 25 mg/L (1:40,000) is safe for sedation for up to 4-8 hours
    - 50-100 mg/L deep narcosis of most small teleost.
    - 300mg/L deep narcosis in larges species.

    Water temperature should be maintained between 5-16 °C. Level 4 narcosis reach within 1-2 min. Recovery within 5 min after removal of fish from solution.

    Large fish like shark or rays should be anesthetized with freshly mixed solution of tricaine in sea water 1 gm/L. Practicable to spray directly in to the mouth or over to the gills. 1 L of the solution may be needed to induce anesthesia within two minutes in a large shark. Recovery is 5-30 minutes from the time fish is re-introduced to untreated seawater.
3. Fresh solution of propoxate (R-7476) in fresh or sea water at 1-4 ppm depending on species.

- 1 ppm anesthetizes rainbow trout to level 4.
- 2-4 ppm large fish such as pike.

The introduction period is 30-60 seconds. The recovery period is 10 min after removal from the anesthesia solution.

4. Use ether at a concentration of 10-15 ml/L of water.

The introduction period for up to level '4' is 3-5 minutes. The recovery period is 5-10 minutes.

Post-anesthetic care of fish: During weaning the fish from anesthesia, it should be placed in a fresh, clean oxygenated water. An assisted ventilation may be necessary for recovery and can be done by rinsing the gill surfaces with fresh, oxygenated water.

Euthanasia: AVMA, 2007, describes overdosage of immobilization drugs as an acceptable method for euthanasia. 5-10 times the anesthetic dose of Tricaine is the most frequently used drug for euthanasia. Injectable agents can also be used for euthanasia. Technique of pouring immersion drugs over the gills can also be done. Ultrasonography or electrocardiography is recommended to confirm asystole followed by administration of additional anesthetic drug or pentobarbitone into the heart or caudal vein.

2.8. Marine Mammals

Marine mammals are a fascinating group of animals that have adapted to live in the ocean. They come in all shapes and sizes, from the giant blue whale, the largest animal on Earth, to the playful sea otter. Despite their dependence on the ocean, marine mammals are all mammals, meaning they share some key characteristics with land mammals. Like us, they breathe air, are warm-blooded, give birth to live young, and nurse them with milk.

2.9. Classification

There are five main groups of marine mammals:

2.10. Cetaceans

This group includes whales, dolphins, and porpoises. Cetaceans are all fully aquatic and have streamlined bodies for swimming. They use echolocation to navigate and find food.

Cetaceans' marine mammals

Of all cetaceans the bottle nosed dolphin (*Tursiops truncates*) has attracted the most in biomedical research. Several factors have been responsible.

i. The respiratory system is quite unique.
ii. The thermoregulatory mechanism necessary to maintain a body temperature of
iii. $36\ ^{0}\mathrm{C}$ in a cold environment are highly specialized.
iv. Drugs commonly used in other species are poorly tolerated or do not work at all.
v. Dolphins are very susceptible to handling stress unless they have been conditioned to the personnel looking after them.

Dolphins breathe through a single nostril on the forehead surface known as a blowhole. This is commonly held closed by a combination of muscular nasal plug and muscular sphincter around the orifice, but it can be opened rapidly whilst the dolphin is swimming at full speed to allow complete respiratory cycle in 0.3 sec. This alone present a challenge to the anesthetist since passage must be anatomically capable of allowing very rapid flow rates to satisfy tidal volume of 5-10 liters of air. However, adult tursiops normally breathe 2-3 times/min inflating thin lungs to about 80% of their capacity (5-10 liters) and holding an apneustic plateau for 20-30 sec before a rapid exhalation. They usually hyperventilate with 4-5 rapid breaths before diving. The larynx is specialized to allow a direct passage between the internal nares and lungs. The trachea is short of wild boar and rigidly constructed of semi-fused cartilaginous rings and the bronchi and bronchioles are also strongly supported by cartilage. An elaborate system of membranes air sacs and tubes open into the nasal passage, but are unlikely to affect anesthetic management. Dolphins maintain a stable body temperature in their cold and highly conductive aquatic environment through several evolutionary adaptations.

i. Their high metabolic rate may be up to 3 times higher in some species than terrestrial mammals of similar weight.
ii. They are equipped with a counter-current heat exchange system in peripheral blood vessel and with a thick insulating layer of blubber.

2.10.1. Recommended anesthetic techniques

Dolphins may be lightly tranquilized with 0.2 mg/kg of diazepam given either orally in fish or I/M after netting but must be carefully observed for respiratory depression. They should then be strapped to a board or canvas stretcher and water at 25-30 ^{0}C sprayed or sponged over their surface to keep them moist. After intubation, induction with the thiopentone (10 mg/kg) injected into the central vein of the flukes. As the jaws relax, they are held apart by attendant and after the glottis has been dislocated from its position within the internal nares, a 24-30 mm cuffed endotracheal tube is introduced directly into the glottis and the cuff is inflated. Positive pressure ventilation with a suitable modified respiratory delivery, 2% halothane in air initially and anesthesia can be maintained there after 0.5-1% halothane in air and oxygen (60:40) gas mixture is supplied. Throughout recovery, until the blow-hole reflex returns (15-45 min after the dolphin has regained consciousness) endotracheal tube can be withdrawn. The glottis relocated within the nares and the dolphin returned to the water as soon as spontaneous respiration was established.

2.10.2. Reflexes that may be observed for depth of anesthesia in the porpoise

(As per Ridgway (1968) taken from the book "Veterinary Anesthesia" Willium V. Lumb, 1984)

i. Contraction of the eyelid on tapping the inner canthus of the eye.
ii. Contraction of the eye muscles when the cornea is touched
iii. Contraction of throat muscles when the hand is inserted into the pharynx.
iv. Retraction of the tongue on being pulled.
v. Reflex movements of the body when the anus is distended
vi. Tail movements.
vii. Movements of the pectoral flippers in response to scratch or pinprick of the chest or axillary region
viii. Movement of blow hole on insertion of the finger into the nares or vestibular sacs.

ix. Vaginal or penile movements when the vagina or prepuce is distended by insertion of the fingers or instruments.

Three of which are particularly valuable during induction. Once the palpebral reflex (blink when the inner canthus is touched), corneal reflex (blink when the cornea is touched) and swimming movement of the free tail flukes each disappear. The depth of anesthesia is adequate for surgery. Anesthesia may be maintained for several hours with a low concentration (0.5-1%) of halothane provided the subject is sufficiently ventilated. The lightest level of anesthesia sufficient to inhibit movement of the tail flukes should be maintained. Porpoises in water are near neutral in buoyancy. Out of water there are abnormal stresses that must be overcome in breaking and maintaining circulation. For this reason, the animal should be returned to the water as soon as possible.

2.11. Pinnipedia

This is a group of fin-footed, carnivorous mammals that live in the ocean and on land. They are widely distributed throughout the ocean and some larger lakes, primarily in colder waters. Pinnipeds range in size from the 1.1 m (3 ft 7 in) and 50 kg (110 lb) Baikal seal to the 6 m (20 ft) and 3,700 kg (8,200 lb) male southern elephant seal, which is also the largest member of Carnivora. This group includes seals, sea lions, and walruses. Pinnipeds have flippers that they use to swim in the water and haul themselves out onto land.

Pinnipeds marine mammals

Pinnipeds share some common characteristics, including:

- Flippers for swimming and moving on land
- Streamlined bodies for swimming
- Thick fur or blubber for insulation
- Keen eyesight and hearing

Pinnipeds are an important part of the marine ecosystem. They prey on a variety of fish, squid, and other marine animals, and they help to keep prey populations in check. Pinnipeds are also a source of food for some indigenous cultures. However, pinnipeds face a number of threats, including habitat loss, pollution, and climate change. Climate change is causing sea ice to melt, which is making it difficult for some pinnipeds to find food and raise their young. Pollution can also harm pinnipeds, by poisoning them or making it difficult for them to find food. It is important to protect pinnipeds so that they can continue to thrive in the ocean for generations to come.

Several important modifications from the normal mammalian respiratory and cardiovascular system are associated with their adaptation to a marine environment and ability to drive to considerable depths. These are of particular interest to the anesthetists. External nares are equipped with sphincters that are close to prevent water from entering the respiratory tract during deep diving. Both seals and sea lions commonly take several deep breaths within a minute and then may hold their breath for 3-5 minutes before another ventilatory phase. This pattern is normal during sleep and may also be observed under deep anesthesia. The cardiovascular system is also adapted so that when the animal dives or when it is frightened, blood is held back in a series or various sinuses until a breath is taken. It then returns to the heart and, if it contains anesthetic agents, a high concentration may suddenly be pumped to the central nervous system.

2.11.1. Anesthetic techniques

Geraei (1973) used ketamine hydrochloride at the dose of 4.5-11 mg/kg of body weight to anesthetize sea lions and seals. Contraindicated in seals that are in poor condition. Induction was 10 mts, recovered within 65 minutes.

In the ringed seal study, ketamine was shown to have a number of desirable features.

- It provides rapid induction of anesthesia
- Threefold margin of safety
- Rapid quiet recovery and uncomplicated
- Can be administered intramuscular which eliminate the need for more handling of aggressive seal.

For these reasons, ketamine meets most of the requirement of an ideal immobilized agent for use under field conditions in seals and sea-lions.

Engelchardt (1977) used Ketamine HCl in young harp seals at the rate of 0.52-7.5 mg/kg body weight (*Phoca groenlandica*). At 60 sec of post injection the seal showed disappearance of vocalization and tactile response. At 20 min of post injection the seal showed a response to tactile stimuli and at 23 min avoided handling when touched and would bite. Joseph *et al.* (1981) used a combination of ketamine and diazepam intramuscular or intravenous at dosages of 1.5 mg/kg and 0.05 mg/kg, respectively. Experimentally induced pneumonia did not after the effects of the drug. But ketamine singly causes muscle tremors and rarely convulsion and death. Introduction and recovery were smoother than with ketamine used alone.

2.11.2. Recommended anesthetic techniques

By observing the work of above-mentioned workers best anesthetic technique for pinnipeds is a combination of ketamine and diazepam to avoid the untoward effect of ketamine alone, at the rate of ketamine (1-2 mg/kg, intramuscular) with diazepam (0.02mg/kg, intramuscular) injected by using 16 gauge needle to ensure blubber penetration. Seals are best caught in nets. First while sea-lion can be injected on land by projectile syringes at close range. After administration of the above combination, anesthesia can also be maintained with halothane delivered by mask or by endotracheal intubation for prolonged surgery.

2.12. Sirenians

This group includes manatees and dugongs. Sirenians are herbivores that graze on seagrass. They have thick blubber to keep them warm in the water. Sirenians, are less likely to require general anaesthesia compared to other marine mammals like dolphins or seals. This is because many procedures can be done with sedation or local anaesthetic, which is less risky for the animal.

Here's a breakdown of anaesthesia in sirenians:

- ***General anaesthesia:*** Rarely used, but may be necessary for major surgery or procedures requiring complete unconsciousness.
- ***Sedation:*** A more common approach, using medications like benzodiazepines to achieve a state of relaxation. This allows for some minor procedures and diagnostic tests.
- ***Local anaesthesia:*** Used to numb a specific area for minor procedures like wound treatment or biopsy.

The reason why sirenians are more suited for sedation or local anaesthesia is due to their unique physiology:

- ***Dive reflex:*** Sirenians, like other marine mammals, have a strong dive reflex that slows down the heart rate and breathing in response to submersion. This can make general anaesthesia with respiratory depression risky.
- ***Respiratory system:*** Sirenians have a less rigid chest wall and smaller airways compared to land mammals. This can make maintaining ventilation during general anaesthesia more challenging.

When anaesthesia is absolutely necessary, veterinarians carefully monitor the animal's vital signs and take steps to minimize risks. This may involve using specialized equipment for ventilation and monitoring body temperature.

Sirenians marine mammals

2.13. Sea otters

Sea otters are the smallest marine mammals. They have thick fur that helps them stay warm in the cold water. Sea otters are expert divers and feed on a variety of prey, including sea urchins and clams.

Sea otters marine mammals

Sea otters, unlike sirenians, can undergo general anaesthesia for necessary medical procedures. Here's what you need to know about anaesthesia in sea otters:

- ***Medications:*** Common anaesthetic combinations for sea otters include opioids like fentanyl along with sedatives like diazepam. The dosage is carefully determined based on the otter's weight and the procedure being performed.
- ***Administration:*** Anaesthesia is typically administered intramuscularly (injection into the muscle) in the hind leg.

- ***Monitoring:*** Veterinarians closely monitor the otter›s vital signs like heart rate, breathing, and body temperature throughout the anaesthesia process.
- ***Risks and considerations:*** Sea otters are susceptible to hypothermia (abnormally low body temperature) under anaesthesia. They also require special attention to maintain proper breathing due to their dependence on frequent air intake while at sea.

Here are some resources for further reading on sea otter anaesthesia:

- Chemical Anaesthesia of Northern Sea Otters (Enhydra lutris): Results of Past Field Studies PubMed: pubmed.ncbi.nlm.nih.gov
- Immobilization of Sea Otters (Enhydra lutris) in Managed Care using Medetomidine-Butorphanol-Midazolam and Dexmedetomidine-Butorphanol-Midazolam VIN: https://www.vin.com/doc/?id=9846873

2.14. Polar bears

Polar bears are the only marine mammals that are considered carnivores. They spend most of their time hunting seals on sea ice. Polar bears are the largest land predators in the world.

Polar bears marine mammals

Marine mammals play an important role in the health of the ocean ecosystem. They help to keep prey populations in check and distribute nutrients throughout the water column. Marine mammals are also a valuable source of food and income for many coastal communities. However, marine mammals face a number of threats, including pollution, habitat loss, and entanglement in fishing gear. It is important to protect these amazing creatures so that they can continue to thrive in the ocean for generations to come.

2.15. Anaesthesia for Crabs and Lobsters

Crabs and Lobsters belongs to class crustacea. After netting, crustaceans are best kept in polythene tubs containing sea water into which oxygen is continuously bubbled. Sand to a depth of 8-10 cm should be supplied on the bottom to allow the animals to burrow. The tubs may be used for transporting them long distance without tranquilization and are also convenient for introducing anesthesia when required.

Crabs and lobsters unlike mammals don't have a complex nervous system and likely don't feel pain in the same way we do. However, there are still standards for humane slaughter of these animals.

There are two main methods for anesthetizing crabs and lobsters:

1. **AQUI-S®** is a commercially available anaesthetic that is specifically designed for crustaceans. It is a mixture of clove oil derivatives and is considered to be a very humane method of anaesthesia.

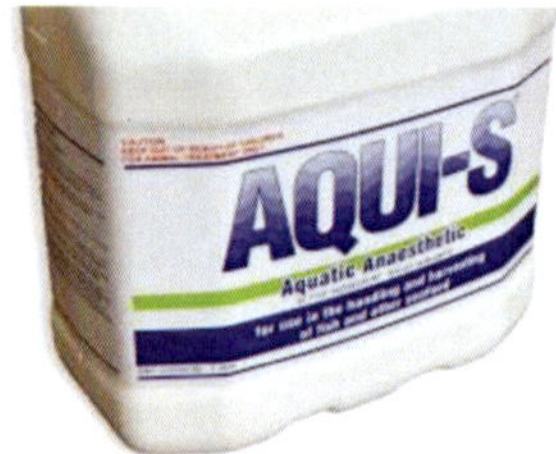

AQUIS® anesthetic for crustaceans

2. **Electrical stunning** - This method involves passing an electric current through the water, which disrupts the nervous system of the crab or lobster. This method is very quick and effective, but it is important to use the correct settings to avoid harming the animal.

   Both of these methods are effective at rendering crabs and lobsters' unconscious before they are slaughtered. Here are some other methods that are sometimes used, but are not generally recommended:

   - **Clove oil** - Clove oil is a natural anaesthetic that can be effective for crabs and lobsters. However, it can be difficult to control the dosage and can be stressful for the animals.
   - **Ethanol** - Ethanol (alcohol) can also be used to anesthetize crabs and lobsters. However, it can be harmful to the animals at high concentrations.
   - Crabs and lobsters have been anesthetized by immersion in solution of isobutyl alcohol and of ether, but tricaine has not proved effective. Isobutyl alcohol in lobsters (Homarus americanus) in concentration of 0.5-14.4 ml/L aerated sea water at 10^{0}C, after brief period of agitation the lobster remained anesthetized for 10 min after removal from the solution and retiring them to sea water.

- Crabs I/V with 0.5 mm (25 gauge) needle inserted through the coxal arthrodial membrane of posterior pereiopod, alphaxolone-alphadolone (30 mg/kg) and procaine (25 mg/kg) were the most effective. Procaine induced anesthesia within 30 sec. of injection and the crab remained asleep for 2-3 hours. The crabs went through a short period of excitement lasting up to 10 sec after injection, but no other disadvantages were reported.

It is important to note that there is still some debate about the best way to anesthetize crabs and lobsters. Some experts believe that all of the methods currently available cause some degree of stress or pain to the animals. More research is needed to develop better methods of anesthesia for these creatures.

2.16. Immobilization of Amphibian

The Class Amphibia includes frogs, toads, salamanders, and caecilians. These fascinating creatures are known for their ability to live in both water and on land. They are ectothermic, meaning they rely on external sources for body heat. Amphibians are an important ecological indicator, meaning their presence or health can reflect the condition of their environment.

Here are the three main orders of amphibians:

1. ***Anura (Frogs and Toads):*** This is the most diverse order of amphibians, with over 8,000 species. Frogs and toads have smooth, moist skin, long legs for jumping, and no tail in adulthood.

Frog

2. ***Caudata (Salamanders):*** Salamanders are typically long and slender with a tail. Unlike frogs, most adult salamanders retain their tails throughout their lives. Some species even have the ability to regenerate lost limbs.

Salamander

3. ***Gymnophiona (Caecilians):*** Caecilians are limbless amphibians that resemble worms or snakes. They live underground and burrow through soil. Caecilians have small eyes or may be completely blind.

Caecilian

There are two main approaches to amphibian immobilization, depending on the context:

1. Research and Veterinary Settings

In these settings, immobilization is crucial for examinations, procedures, or transport. Vets and researchers use various methods to ensure the amphibian's safety and minimize stress. Here are some common techniques:

- **Anesthetics:** Certain medications like lidocaine or chloretone can be used to anesthetize amphibians, depending on the species and procedure. This renders them unconscious and immobile.
- **Cooling:** Amphibians are ectothermic, meaning their body temperature depends on the environment. By carefully lowering the ambient temperature, vets can induce a state of torpor (a dormancy-like state) in some amphibians, reducing movement.

2. Short-term Restraint

For short examinations or handling, some gentler restraint methods might be used:

- **Mesh Bags:** For small amphibians, a breathable mesh bag can be a secure way to contain them while allowing air circulation.
- **Tubes:** For some species, a smooth-sided tube just large enough for the amphibian to fit snugly can be used for short-term restraint.

Important Points:

- Never use harsh chemicals or methods that can harm the amphibian's delicate skin.
- The chosen method should be appropriate for the specific amphibian species, size, and situation.
- Always prioritize the amphibian's well-being and minimize stress during immobilization.

Gases and anesthetic agents in solution are rapidly transferred across the skin of amphibians to diffuse into the circulation, thus providing a simple route for anesthetic administration. Inhalation and water-soluble agents have been administrated by sitting the amphibian in a moist atmosphere into which an anesthetic mixture is passed by immersing the animal in an anesthetic solution or by wrapping it in a cloth moistened with the agent. Alternatively, administration may be by intravascular or intraperitoneal injection. The simplest route for intravascular injection is through the paired dorsal lymph sacs sited one on each side of the last vertebra and usually easily identified by their rhythmic beating. Volumes of up to 3 ml may injected into the animal at any one time.

2.16.1. Recommended anesthetic techniques

Although the esophagus occasionally prolapses into the buccal cavity during anesthesia, vomiting is not a problem in from, so they can be fed freely until anesthetized. No preanesthetic treatments are necessary. In fact, prior administration of chlorpromazine was held to be highly toxic to frogs. They should be acclimating to a water temperature of 20-22 ^{0}C for two days prior to anesthesia. Depth of anesthesia is easily assessed by the withdrawal reflex when digits are pinched and by the response to pin prinks.

Frogs must not be immersed in water during the recovery period or they may drown. However, the entire body surface must be kept moist throughout anesthesia and recovery. Since metabolic rate and hence recovery is enhanced by higher temperatures, particularly in the case of ectotherms, frogs should be allowed to recover quickly in an ambient temperature of 24-26 ^{0}C.

Adult frogs are anesthetized most easily by simple inhalational or immersion techniques. The methods of choice are described in order of preference.

1. Inhalation of methoxyflurane is the simples and safest method. Suitable containers of glass or clear rigid plastic are prepared for introduction by placing a wad of cotton wool in the bottom and spraying methoxyflurane on to it. A total of 0.5-1 ml is sufficient to produce an induction concentration of vapour (3%) in a 1 litter container. The container should then be left for 20-30 min with the lid closed to allow the volatile liquid to vaporize adequately. Deep anesthesia is induced within two minutes of the frogs being placed in the container and is maintained for a further 40 min after they are taken out. Frogs should be allowed to recover in a separate container on moist cotton wool after being washed in warm (24 ^{0}C) tap water. Full recovery takes about 7 h the method is safe and analgesia is excellent.
2. Immersion in a 0.1% tricaine solution at a temperature of 21 ^{0}C is also effective. Introduction of anesthesia occurs in 5-20 min. Anesthesia is maintained by wrapping in a moistened cotton wool with tricaine. The recovery time is 25-75 min. Muscular relaxation and analgesia is both very good. If rapid recovery is essential to the experiment or surgery, solution of tricaine can be injected in the dorsal lymph at 13 mg/kg. Induction takes 3-5 min and recovery is complete within 30 min.
3. If no other agents available ether can be used as an inhalation agent in the same way as described for methoxyflurance, using 1-2 ml in a 1 liter container. Alternatively, it can be used as a 4% solution in water. Induction takes place in 3-4 min and recovery time is 30 to 40 min. Both are relatively short however, ether is highly irritant to frog skin and the animals are stressed when first introduced to it.

   Tadpoles are safety anaesthetized in a 1:3000 aqueous solution of tricaine and induction takes only 60 sec. Recovery is complete in 3 min if the tadpoles are washed in warm tap water and returned to their resident aquarium,

Order-Urodela (Newts, salamanders, mud puppies)

The general management and anesthetic techniques described for frogs and toads applies to Urodeles. Immersion is an aqueous solution of tricaine at a concentration of 1:2000 is the simplest method for adults, while a 1:3000-1:5000 concentration is sufficient for larval stages.

3

Chemical Restraint of Reptile

3.1. Introduction

Various procedures performed on reptiles often necessitate chemical restraint or general anesthesia. Even routine tasks like physical examination and transportation can pose challenges, particularly with large or uncooperative reptilian patients. For safety reasons, venomous species are typically evaluated under chemical restraint. The respiratory anatomy and physiology of reptiles differ significantly from those of more familiar mammals, influencing their response to commonly used anesthetic agents. Historically, anesthetics recommended for reptiles had notable metabolic effects, but inhalant anesthesia has become more popular in recent times. Nonetheless, in species capable of anaerobic metabolism, induction with these agents may not be feasible (Girling 2013).

Reptiles generally experience less stress compared to other counterparts like avian and mammals, making restraint less risky, especially for debilitated animals. However, it's crucial to assess factors that could pose dangers to both the animal and the handler during restraint:

- Is the patient experiencing respiratory distress, where excessive manual manipulation could be hazardous?
- Is the species fragile? For instance, day geckos are exceedingly delicate and may shed their tails when handled. Similarly, species like green iguanas are susceptible to conditions like metabolic bone disease, leading to spontaneous fractures.
- Is the species naturally aggressive? Some, such as snapping turtles, tokay geckos, and rock pythons, exhibit aggressive tendencies.
- Furthermore, it's important to note that many reptile species naturally carry Salmonella spp. in their gut. Therefore, maintaining personal hygiene is crucial when handling these patients to prevent the transmission of zoonotic diseases (Bennett 1996).

3.2. Classification

Snakes belong to the suborder Serpentes within the order Squamata. Squamata is the largest order of reptiles, consisting of over 10,000 species including lizards, amphisbaenians, and of course, snakes.

The family Acrochordidae includes Elephant trunk snake (Acrochordus javanicus).

The family Boidae divided into two subfamilies. The subfamily Boinae includes common boa (*Boa constrictor*) and giant anaconda or green anaconda (*Eunectes murinus*). The subfamily Pythonia includes African rock python (*Python sebae*), Reticulated python (*Python reticulates*) and Indian python (*Python molurus*).

The family Colubridae includes Pine snake (*Pituophis melanoleucus*), Pacific gopher snake (*Pituophis catenifer catenifer*), Bull snake (*Pituophis catenifer sayi*), Black rat snake (*Pantherophis obsoletus*), Garter snake (*Thamnophis elegans terrestris*), Ring neck snake (*Diadophis punctatus*), Tiger rat snake (*Spilotes pullatus*) and East Indian water snake (*Homalopsis buccata*).

The family Elapidae includes Black cobra (*Naja melanoleuca*), Spitting cobra (*Naja nigricollis*), Egytpian cobra (*Naja haja*) and Cape cobra (*Naja nivea*).

The family Viperidae have two subfamilies. The subfamily Crotalinae includes timber rattle snake (*Crotalus horridus*), cane break rattle snake (*Crotalus horridus atricaudatus*), red diamond black rattle snake (*Crotalus ruber*), Korean viper (*Gloydius brevicauda*) and white-lipped tree viper (*Trimeresurus albolabris*). The subfamily Viperinae includes Russell's viper ***(Daboia siamensis)*,** Palestine viper (*Daboia palaestinae*), puff adder (*Bitis arietans*), rhinoceros viper (*Bitis nasicornis*) and gaboon viper (*Bitis gabonica*)

Squamata Order Reptile

Here are some key characteristics of Squamata:

- Scaly skin
- Claws (except for snakes and some legless lizards)
- Eardrums (except for some lizards and snakes)
- Three-chambered heart (except for some lizards)

Snakes are easily distinguished from other squamates by their lack of limbs and external ears. They also have a long, forked tongue that they use to smell and taste their surroundings. Snakes come in a wide variety of shapes and sizes, from tiny blind snakes that are just a few inches long to giant constrictors that can grow over 30 feet in length.

3.2. Anatomy and Physiology of Reptiles

3.2.1. Cardiopulmonary Anatomy

a) **Glottis Position:**

Squamates (Lizards and Snakes):

- Location: Rostral, at the base of the tongue.
- Function: Positioned to aid in breathing while holding prey (Bennett 1998).

Chelonians (Turtles and Tortoises):

- Location: Caudal, in the throat region.
- Function: Positioned further back to facilitate feeding and breathing given their unique head retraction mechanism (Scarabelli *et al.*, 2022).

Crocodilians (Alligators and Crocodiles):

- Structure: Covered by an epiglottal fold, known as the basihyal valve.
- Function: Allows them to open their mouths underwater without ingesting water, aiding in their semi-aquatic lifestyle (Fleming 2001).

b) **Tracheal Structure:**

Squamates:

Rings: Incomplete tracheal rings.

Significance: Provides flexibility to accommodate the elongation and movement of the body.

Chelonians and Crocodilians:

- **Rings:** Complete tracheal rings.
- **Significance:** Provides structural rigidity and support, crucial for maintaining an open airway during head retraction (chelonians) and during vigorous movements (crocodilians).

c) **Lung Structure:**

- **Squamate Reptiles:** Simple sac-like structure with endothelial lining.
- **Chelonians:** Complex lung with invaginations.
- **Crocodilians:** Highly developed lungs, resembling mammalian lungs.

Snakes: Varied lung development, with some species having vestigial left lungs and functional air sacs.

3.2.2. Respiratory Mechanisms

a. **Breathing Methods:**

- Mouth Breathing Capabilities: Many reptiles can breathe through their mouths if necessary, though it is not their primary method.
- Primary Nasal Breathing: Most reptiles primarily breathe through their nostrils, which helps filter and humidify the air (Bennett 1998).

b. **Muscle Movements:**

- Negative Pressure for Inspiration: Generated by respiratory muscle movement, such as the contraction of the diaphragm in some reptiles.
- Smooth Muscle Contraction: Aids in the movement of air within the lungs and airways, contributing to ventilation efficiency.

c. **Ventilation Assistance:**

- Importance During Anesthesia: Assisted ventilation is crucial as anesthesia can paralyze respiratory muscles, leading to respiratory failure without intervention.
- Paralysis of Respiratory and Skeletal Muscles: During anesthesia, reptiles may lose the ability to breathe independently, necessitating mechanical ventilation to maintain adequate oxygenation.

3.2.3. Metabolic Adaptations

Anaerobic Metabolism:

- Adaptation: Reptiles have the ability to switch to anaerobic metabolism in low-oxygen environments.
- Survival: This adaptation allows certain species to survive in oxygen-deprived conditions for extended periods.
- Challenges in Anesthesia Induction: Reptiles' metabolic adaptations can make anesthesia induction challenging, as their ability to switch to anaerobic metabolism can alter their response to anesthetic agents.

3.3. Pre-Anesthetic Preparation

Pre-anesthetic preparation is a crucial process to ensure the safety and effectiveness of anesthesia for a surgical or medical procedure. Here are the key steps typically involved (Girling, 2013)

3.3.1. Weight Measurement

- Accurate weight measurement is crucial, especially for small reptiles.
- Scales accurate to 1 gram are recommended for smaller reptiles to ensure precise dosage calculation.

3.3.2. Blood Testing

It's advisable to conduct biochemical and hematological tests before administering chemical immobilizing drugs. Blood samples can be obtained from:

- Jugular vein in Chelonia
- Dorsal tail vein in Chelonia
- Ventral tail vein in snakes, Crocodylia, and lizards
- Palatine vein or cardiac puncture in snakes (although sedation or anesthesia may be required for blood collection from these routes).

3.3.3. Fasting

- Snakes require fasting to prevent regurgitation and pressure on the lungs or heart.
- It's recommended to ensure no prey has been offered in the two days prior to anesthesia.
- Other reptiles, such as chelonians, require less fasting, as they rarely regurgitate.
- Live prey should not be fed to insectivores like leopard geckos 24 hours prior to anesthesia, as the prey may still be alive when the reptile is anesthetized.

3.4. Routes of Drug Administration: Administering drugs to reptiles requires understanding their unique anatomy and physiology. Here are the main routes of drug administration in reptiles (Bennett1998)

3.4.1. Intramuscular (IM) Administration

- Most injectable anesthetic drugs are administered IM.
- Oral administration is unreliable due to unknown absorption and distribution.

- Caution is needed with drugs that may be toxic to or eliminated by the kidneys, avoiding injection in the caudal half of the body.
- Snakes: Front third of the body length is preferred.
- Provides straightforward and less invasive drug delivery.
- Suitable for various clinical situations in snake medicine.
- Large lizards and crocodiles: Upper leg muscles.
- Small lizards: Hind limb muscles.
- Chelonians: Quadriceps muscle group on hind legs or muscle masses of forelegs.

3.4.2. Subcutaneous (SC) Administration

- Similar to mammals, SC injections may be administered.

3.4.3. Intracoelomic (IC) Administration

- Rapidly absorbed route.
- Needle inserted in the left caudolateral abdomen, directed away from viscera, with aspiration before injection.

3.4.4. Intravenous (IV) Administration

- Difficult in most species.
- In chelonians, the right or left jugular vein may be used.
- Risk of leakage from dorsal tail vein or brachial venous plexus.
- Ventral abdominal vein and ventral tail vein used for drug administration in lizard.
- Cephalic vein accessible in some species with surgical cut-down in lizard.

3.4.5. Intraperitoneal (IP) Administration

- Snakes, lizards, and crocodiles: Mid-ventral approach to access the peritoneal cavity.
- Chelonians: Administer injections into the fossa anterior to the hind leg.

3.4.6. Intracardiac Administration

- Swiftest method for drug delivery and blood sampling in snakes.
- Heart's topographic position on the ventral body wall facilitates identification.
- Caution advised by McFadden *et al.,* (20011) due to stress and risk of myocardial damage.
- Requires careful consideration and expertise before use.

3.5. Handling and restrain of reptiles

Reptiles generally experience less stress from handling compared to birds, but restraint still requires careful consideration to avoid potential dangers to both the animal and the handler (Girling, 2013).

3.5.1. Serpents/ Snake handling and restraint

Snakes are easily distinguished from other squamates by their lack of limbs and external ears. They also have a long, forked tongue that they use to smell and taste their surroundings. Snakes come in a wide variety of shapes and sizes, from tiny blind snakes that are just a few inches long to giant constrictors that can grow over 30 feet in length.

Reptiles generally good subjects for anesthesia but show several features that may prove unfamiliar to the clinical accustomed to working with mammals. Snakes pose particular anesthetic problems. Their ectothermic nature and hence variation in their response to anesthetic agents at different temperatures. Basal metabolic rate is low and directly related to environmental temperature. The relative resistance to hypoxia (especially certain aquatic chelonians) and ability of many species to hold their breath for several minutes to further delay induction with inhalational agents. Anatomical features also important. The single lung consists of an elongated thin walled hallow tubular organ. The trachea is unusual in that it is open along the side within the lung. The lung terminates in an indistinguishable air sac (similar to that seen in birds) which may continue almost the cloaca. Without a diaphragm, an inspired breath can inflate the snake for almost its entire length. This is especially obvious in certain species such as hognose snake. Caldrewood (1971) has suggested that these features predispose to respiratory arrest.

Some clinical consideration that the reptile should be examined, before hand and must not have eaten during previous 18 hours (lizards and chelonians), 96 hours (snakes and crocodiles). There is no need of withholding the water. Assessment of depth of anesthesia is not always easy in reptiles in view of the paucity or reliable reflexes. For example the corneal and palpebral reflexes are non-existent in snakes and certain lizards and are little use in other groups. The righting reflex of most values and should be coupled with close observation of respiration and response to painful stimuli.

Manual Restraint for Handling

- Handling larger constricting species of snakes, especially those over 10 feet in length, requires careful consideration and adherence to safe operating practices. The "buddy system" where snakes longer than 5–6 feet are handled by two or more people, is essential for safety. This ensures that if one handler becomes ensnared, others can assist in disentangling them by unwinding from the tail end first.

- Maintaining a firm but gentle grip is crucial to avoid injuring the snake and preventing the release of myoglobin from muscle cells, which can damage the kidneys' filtration system. This emphasizes the importance of proper handling techniques that prioritize the well-being of both the handlers and the snakes. By following these guidelines and working together as a team, handlers can minimize risks and ensure the safe handling of larger constricting snake species.
- When dealing with venomous or very aggressive species, using snake hooks and tongs can provide a safer means of initial restraint (Fig 1). These hooks and tongs allow handlers to manipulate the snake at a safe distance, minimizing the risk of bites or strikes. Proper use of snake hooks involves looping them under the body of the snake to move it into a container or trap the head against the floor before grasping it with the hand. It's crucial to note specific precautions for dealing with certain venomous species, such as the spitting cobra family, where handlers should wear protective gear like plastic goggles or a face visor to prevent venom from reaching their eyes and mucous membranes.

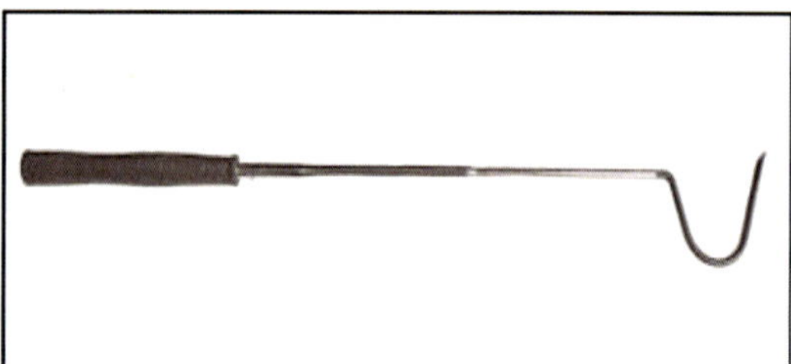

Snake Hooks

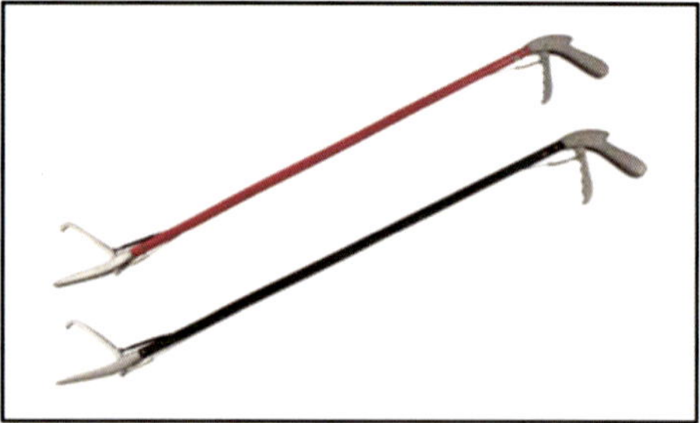

Snake tongs

3.5.2. Handling of Lizards

Handling lizards requires careful consideration of their anatomy and behavior to ensure the safety of both the handler and the animal. Here are some guidelines for safely handling lizards (Ferreira and Mans 2022).

- **Identify danger areas:** Lizards main danger areas to handlers are their claws, teeth, and in some species, such as iguanas and monitor lizards, their tails, which can lash out in a whiplike fashion.
- **Understanding behavior:** Different lizard species exhibit varying levels of docility and aggression. While geckos and bearded dragons are generally docile, species like green iguanas may display aggression, especially sexually mature males. They may also react more aggressively towards female owners and handlers due to their ability to detect pheromones secreted during the menstrual cycle.
- **Proper restraint:** When handling lizards, it's essential to grasp them securely to prevent injury to both the handler and the animal. Grasping around the shoulders (the pectoral girdle) and the pelvic girdle from the dorsal aspect allows for control of the limbs while avoiding the danger areas. Handlers should allow some flexibility as lizards may struggle, and overly rigid restraint could damage the spine.

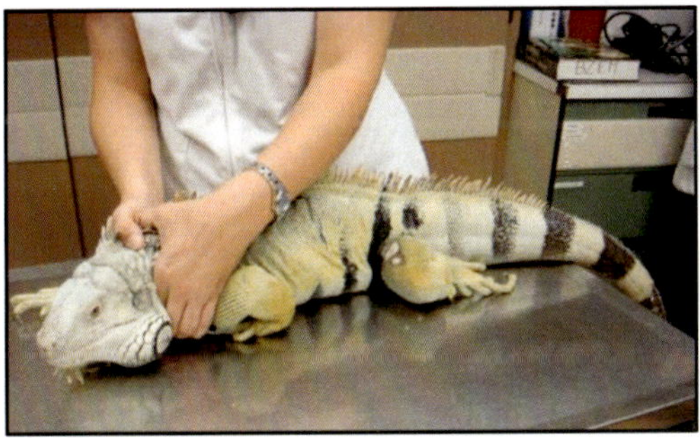

Handling of lizards during examination

- **Handling aggressive lizards:** More aggressive lizards, such as some iguanas, may need to be pinned down initially. Using a thick towel to control the tail and claws can be helpful in such cases. Gauntlets may be necessary for particularly aggressive or large lizards, or those with a venomous bite.
- **Avoiding over-restraint:** It's important not to use too much force when restraining lizards, especially those with skeletal problems like metabolic bone disease. Lizards lack a diaphragm, so overzealous restraint can lead to increasing pressure on the lungs.

- **Handling fragile species:** Fragile lizard species, such as day geckos, are best examined in a clear plastic container. Latex-free gloves and soft cloths should be used for examination to avoid damaging their skin.
- **Preventing autotomy:** Lizards should never be restrained by their tails, as many species may shed their tails as a defense mechanism. However, not all lizards will regrow their tails, and some may be left tailless, especially as they age. It's essential to be mindful of this when handling lizards to prevent unnecessary stress or injury.

3.5.3. Handling of Chelonia

Handling techniques for different species of Chelonia vary based on factors such as temperament and physical characteristics. Here's a breakdown:

- **Harmless Chelonia (e.g., Mediterranean tortoises)**: These species are generally docile but possess surprising strength. They can be safely held with both hands, one on either side of the main part of the shell behind the front legs. To keep them still for examination, they can be placed onto a cylindrical object or stack of tins, raising their legs clear of the table surface as they balance on the center of the underside of the shell (plastron).

Handling of mediterranean tortoises

- **Aggressive Chelonia (e.g., snapping turtles and alligator snapping turtles):** Handling aggressive species requires caution to avoid being bitten. It's essential to firmly hold the shell on both sides behind and above the rear legs, keeping a safe distance from the head. Chemical restraint may be necessary to examine the head region safely in these species.

Handling of snapping turtles

- **Soft-shelled and aquatic Chelonia:** For species with delicate shells or those adapted to aquatic environments, such as soft-shelled turtles, handling should be done with care to prevent shell damage. Soft cloths and latex-free gloves can be used to provide protection while minimizing stress for the turtle.

3.5.4. Handling of Crocodilians

- Handling crocodilians, whether for research, conservation, or management purposes, requires great caution and expertise due to their formidable jaws and powerful bodies. Small specimens can be restrained by carefully grasping the base of the tail while securing the snout closed with a rope halter or noose. It's essential to understand that crocodilians' jaw muscles are designed primarily for closing, not opening, which allows for relatively fine rope or tape to keep their mouths shut.

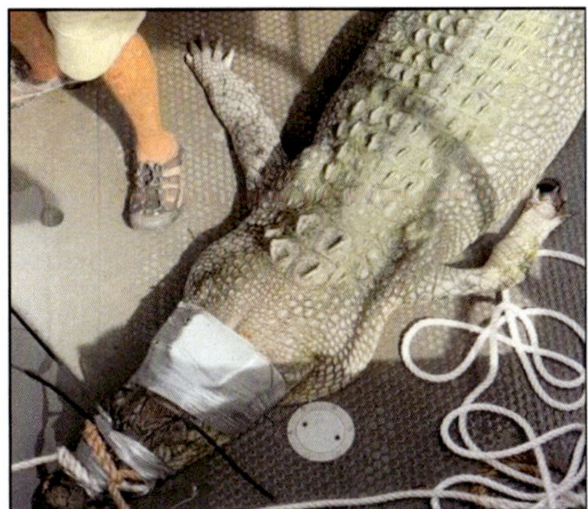

Handling of crocodile

- Approaching them head-on is advisable, considering their poor binocular vision, though members of the alligator family have slightly better vision. Close proximity to a crocodilian demands acute awareness of both head and tail movements, as both can pose significant threats simultaneously.

- When dealing with larger specimens, it's wise to employ teams of people equipped with nets, snout snares, and other restraining tools to swiftly and safely manage the animal. In some cases, chemical immobilization via dart guns may be necessary for the safety of both the handlers and the crocodilian. Such procedures underscore the importance of meticulous planning, proper equipment, and well-trained personnel when interacting with these powerful reptiles.

3.6. Chemical restraint for reptiles

3.6.1. Pre-anesthetic Medications for Reptiles

Pre-anaesthetic medications are essential in veterinary practice for ensuring cardiopulmonary and central nervous system stabilization, smooth anaesthetic induction, muscle relaxation, analgesia, and sedation. Here, we discuss several classes of premedications used in reptiles, including their dosages and effects (Girling 2013).

3.6.1.1. Antimuscarinic Medications

Atropine (0.01–0.04 mg/kg IM) and glycopyrrolate (0.01 mg/kg IM) can be used to reduce oral secretions and bradycardia. However, these concerns are typically less significant in reptiles. Notably, antimuscarinics can increase the thickness of mucus secretions, potentially leading to airway blockages. In herbivorous reptiles, these medications may also cause prolonged periods of ileus (Malley 1997).

3.6.1.2. Tranquilizers

Acepromazine (0.1–0.5 mg/kg IM) is administered approximately one hour before anaesthetic induction to lower the required anaesthetic dose. Alternatives include diazepam (0.22–0.62 mg/kg IM in alligators) and midazolam (2 mg/kg IM in turtles), which also serve to reduce the necessary levels of anaesthetic (Bennett, 1998).

3.6.1.3. Alpha-2 Adrenoceptor Stimulants

Xylazine at 1 mg/kg, given 30 minutes prior to ketamine, can decrease the ketamine dose needed in crocodilians. Medetomidine, administered at 100–150 mg/kg, significantly lowers the ketamine dose required in chelonians and can be reversed with atipamezole at 500–750 mg/kg. Both medications can cause a drop in blood pressure and cardiac output, necessitating cautious use in debilitated reptiles (Lawton 1992).

3.6.1.4. Opioids

Butorphanol (0.4 mg/kg IM), given 20 minutes before anaesthesia, provides analgesia and reduces the necessary anaesthetic amount. It can be combined with midazolam at 2 mg/kg. Butorphanol does not have sedative or general anaesthetic properties but exhibits anaesthetic-sparing effects, allowing for lower levels of gaseous anaesthetic to be used (Bennett 1996).

3.6.1.5. Fluids

Fluid therapy is a critical component of veterinary care for reptiles, especially when they are undergoing surgery or are ill. Proper hydration is essential for maintaining physiological functions, promoting recovery, and ensuring the effectiveness of medications (Frye, 1991). Maintenance fluid levels in reptile patients are typically 25–30 mL/kg/day. Fluid Types are:

- Isotonic solutions (e.g., lactated Ringer's solution, normal saline) are commonly used to correct dehydration.
- Electrolyte solutions may be needed if there are specific imbalances detected in blood tests.

3.6.2. Induction of Anesthesia in Reptiles

General Considerations

It is crucial to note that reptiles should never be immobilized by chilling or cooling, as this method does not provide analgesia and has serious welfare implications.

3.6.2.1. Dissociative Anaesthetics

Ketamine: Ketamine is a widely used dissociative anesthetic in veterinary medicine, including for reptiles (Bennett 1996). Its unique properties make it useful for various procedures requiring anesthesia or sedation. Some important points to be consider are:

- Sedation Dose: 22–44 mg/kg IM
- Surgical Anaesthesia Dose: 55–88 mg/kg IM
- Premedicant Dose Reduction: Lower doses needed if combined with midazolam or medetomidine.
- Maximum Safe Dose: Exceeding 110 mg/kg can cause profound bradycardia and death.
- Onset: Effects are seen in 10–30 minutes.
- Duration: Can last up to 4 days, particularly at low environmental temperatures.

- Usage: Primarily used at lower doses for sedation, facilitating intubation, and maintenance of gaseous anaesthesia in species like chelonians.
- Pain on Administration: Can be painful when administered.
- Excretion: Administer in the cranial half of the body to avoid rapid excretion via kidneys.Renal Disease Consideration: Prolonged recovery in reptiles with renal disease.
- Combination Dosage:
- Small Reptiles (<2 kg): 5 mg/kg ketamine with 100 μg/kg medetomidine.
- Larger Reptiles (>2 kg): 7.5 mg/kg ketamine with 75 μg/kg medetomidine.

3.6.2.2. Alfaxalone

- Dosage: 9 mg/kg
- Onset (IV): Allows intubation within 3–5 minutes.
- Onset (IM): Takes 25–40 minutes for induction.
- Effects: Primarily used for intubation and immobilisation, rather than full anaesthesia (Sheelings *et al.*, 2010).

3.6.2.3. Propofol

- **Advantages:** Rapid induction and recovery, short elimination half-life, minimal organ metabolism.
- **Disadvantages:** Requires intravenous access, can cause transient apnoea and cardiac depression.

Doses:

- Green Iguanas (IO Route): 10 mg/kg.
- Chelonians (IV via Dorsal Coccygeal Vein): 10–15 mg/kg for induction in under 1 minute.
- Anaesthesia Duration: Provides 20–30 minutes of anaesthesia.
- Giant Chelonia: Lower doses of 1–2 mg/kg IV/IO used.

Clinical Applications:

- **Intubation:** Both alfaxalone and propofol are used to facilitate intubation.
- **Anaesthesia Maintenance:** Ketamine and propofol are used in combination with gaseous anaesthetics for maintaining anaesthesia.
- **Special Considerations:** Each agent has specific benefits and limitations depending on the species, health status, and procedural requirements.

3.6.2.4. Depolarizing Muscle Relaxants

Depolarising muscle relaxants are used in veterinary practice to achieve muscle relaxation and immobilization. However, it is important to note that these agents do not provide analgesia and should therefore be used in conjunction with other forms of anaesthesia or for specific purposes such as transportation or intubation.

a. **Succinylcholine:** Succinylcholine is a neuromuscular blocking agent that provides immobilization without analgesia. Key points regarding its use include:
   - **Purpose:** Should only be used to facilitate the administration of another anaesthetic or for transportation, not as a sole anaesthetic agent.
   - **Metabolism:** Recovery is dependent on liver metabolism, so its use should be avoided in animals with potential liver disease.
   - **Dosage and Effects:**
   - Giant Chelonia: 0.5–1 mg/kg IM. Allows intubation and conversion to gaseous anaesthesia.
   - Crocodilians: 3–5 mg/kg IM. Immobilization occurs within 4 minutes, with recovery in 7–9 hours.
   - **Respiration:** Respiration usually continues unassisted at these doses, but assisted ventilation facilities should be available as respiratory muscle paralysis can occur.
   - **Reversal:** Succinylcholine cannot be reversed; the patient must be ventilated until the drug is excreted.

b. **Gallamine:** Gallamine is another muscle relaxant with different properties and uses:
   - **Purpose:** Used to achieve immobility with the advantage of being reversible.
   - **Dosage and Effects:**
   - Crocodiles: 0.3–1.5 mg/kg. Immobility achieved in 15–30 minutes, with a recovery time of 1.5–3 hours.
   - **Reversal:** Gallamine can be reversed with neostigmine at a dose of 0.25 mg/kg.
   - **Clinical Considerations**
   - **Safety:** Always ensure that appropriate ventilation support is available when using muscle relaxants, particularly those that cannot be reversed.

- **Combination with Analgesia:** Since these agents do not provide analgesia, it is crucial to combine their use with adequate analgesic and anaesthetic agents.
- **Species-Specific Dosage:** Adjust dosages and protocols based on the species, size, and health status of the reptile.
- **Monitoring:** Continuous monitoring of vital signs and readiness to provide respiratory support are essential when using these drugs.

3.6.2.5. Gaseous Anaesthetics

Gaseous anaesthetics are widely used in veterinary practice, particularly for their ease of administration and control over anaesthesia depth. However, they come with their own set of advantages and disadvantages.

Advantages

- **Ease of Administration via Face Mask:** Gaseous anaesthetics can be easily administered using a face mask, gas chamber, making it convenient for induction and adjustments during procedures.

Gaseous anesthetics administered via Face Mask

- **Pain-Free:** The administration of gaseous anaesthetics is generally pain-free, providing a more comfortable experience for the patient compared to injectable agents.
- **Minimal Tissue Trauma:** Since gaseous anaesthetics do not require injection, there is minimal risk of tissue trauma or necrosis at the administration site.

Disadvantages

- **Breath-Holding:** Breath-holding, particularly in species such as chelonians, can be a significant issue during the induction of gaseous anaesthesia, complicating the process.
- **Environmental Pollution:** The use of gaseous anaesthetics can contribute to environmental pollution if not properly managed and scavenged.

- **Health Risk to Anaesthetist:** There is a health risk to the anaesthetist due to potential exposure to anaesthetic gases, necessitating adequate ventilation and scavenging systems in the workplace.
- **Risk with Dangerous Reptiles During Handling:** Handling dangerous reptiles for face mask administration can pose significant risks to the veterinary staff, requiring additional precautions and expertise.

3.6.3. Maintenance of Anaesthesia

Maintaining anaesthesia in reptiles can be achieved through the use of injectable or gaseous agents. Each type has its specific protocols, benefits, and challenges. Below is an overview of the use of injectable agents for maintaining anaesthesia.

3.6.3.1. Injectable Agents

3.6.3.1.1. Dissociative Anaesthetics (Ketamine)

Dosage: 55–88 mg/kg IM for anaesthesia.

- Note: Higher doses increase recovery time, potentially extending to several days.
- Doses above 110 mg/kg can cause respiratory arrest and bradycardia.

Combinations:

- With Midazolam: 2 mg/kg IM midazolam with 40 mg/kg ketamine in turtles (Bennett, 1996).
- With Xylazine: 1 mg/kg IM xylazine, administered 30 minutes prior to 20 mg/kg ketamine in large crocodiles (Lawton, 1992).
- With Medetomidine: 100 mg/kg IM medetomidine with 50 mg/kg ketamine in kingsnakes (Malley, 1997).

3.6.3.1.2. Propofol

Usage: Provides 20–30 minutes of anaesthesia, suitable for minor procedures like wound repair, intraosseous or intravenous catheter placement, or oesophagostomy tube placement.

Dosage: Can be topped up at 1 mg/kg/min IV/IO.

- Apnoea is extremely common; thus, intubation and ventilation with 100% oxygen are required.

3.6.3.1.3. Alfaxalone

Usage: Suitable for induction and short periods of anaesthesia (average 25 minutes).

Dosage: 9 mg/kg IV/IO.

- Topping up is possible but may not provide adequate anaesthesia for invasive procedures in all species.
- Higher doses (e.g., 30 mg/kg in green iguanas) may be required for surgical anaesthesia but can cause apnoea (Kischinovsky and Bertelsen, 2011).

Recovery: Recovery times typically range from 1 to 4 hours.

Clinical Considerations

- **Combining Agents:** Using combinations of ketamine with other agents like midazolam, xylazine, or medetomidine can enhance the depth of anaesthesia and reduce the required dose of each drug, minimizing potential side effects.
- **Monitoring:** Continuous monitoring of respiratory and cardiovascular functions is crucial, especially when using agents that can cause apnoea or cardiovascular depression.
- **Ventilation Support:** Be prepared to provide ventilation support when using drugs like propofol or high doses of alfaxalone, which can cause significant respiratory depression.
- **Recovery Management:** Plan for prolonged recovery periods, especially with higher doses of injectable agents, and ensure the reptile is kept in an appropriate environment during this time.

3.6.3.2. Gaseous Agents

Gaseous anaesthetics are commonly used for the maintenance of anaesthesia due to their ease of control and rapid recovery. Below is an overview of the main gaseous agents used in reptilian anaesthesia, their advantages, disadvantages, and specific usage considerations.

3.6.3.2.1. Isoflurane

Isoflurane is the preferred gaseous maintenance anaesthetic for reptiles due to its favorable properties.

Metabolism and Solubility

- Minimally metabolised in the body (0.3%).
- Very low blood–gas partition coefficient (1.4 compared to 2.3 for halothane), leading to low solubility in blood.
- Low fat solubility, meaning it is not stored in the body.

Recovery: Rapid recovery as the drug is quickly excreted from the lungs once administration stops.

Properties: Excellent muscle relaxation and good analgesia during anaesthesia.

Safety: Apnoea precedes cardiac arrest, unlike halothane, making it safer in terms of monitoring respiratory function.

Usage

- Induction: 4–5% in an induction chamber for species not prone to breath-holding.
- Maintenance: 2–3% via an endotracheal tube, depending on the procedure.
- Adaptations: Long, thin face masks made from 20 mL and 60 mL syringes can be used to induce snakes.

3.6.3.2.2. Sevoflurane

Sevoflurane is another safe gaseous anaesthetic for reptiles, though it has some specific considerations.

Solubility: Highly insoluble in the bloodstream, requiring increased ventilation rates (above the usual 4–6 breaths per minute) to maintain anaesthesia.

Limitations: May not be suitable as the sole maintenance agent for certain species (e.g., leopard geckos, bearded dragons, green tree pythons) as they may wake up once the induction agent wears off.

Usage:

- Induction: Higher levels than isoflurane, typically 6–8%.
- Maintenance: 3–4%.

3.6.3.2.3. Nitrous Oxide

Nitrous oxide can be used in combination with other gaseous anaesthetics like isoflurane.

Advantages

- Reduces the percentage of gaseous anaesthetic required for both induction and maintenance.
- Provides good muscle relaxation and excellent analgesia, particularly useful in orthopaedic procedures.

Disadvantages

- Tendency to accumulate in hollow organs, problematic for herbivorous reptiles with capacious hindguts.

- Requires organ metabolism for full excretion, posing risks for patients with serious disease.
- Prolongs anaesthetic recovery times by up to 50%.

Clinical Considerations

- **Combining Agents:** Nitrous oxide is often used in combination with isoflurane to enhance muscle relaxation and analgesia while reducing the required dose of the primary anaesthetic.
- **Species-Specific Responses:** Some reptiles may not respond well to certain gaseous agents, necessitating adjustments in anaesthetic protocols.
- **Ventilation Support:** Ensure adequate ventilation support, especially when using agents like sevoflurane that may require higher ventilation rates.
- **Monitoring and Safety:** Continuous monitoring of respiratory and cardiovascular functions is critical to ensure the safety of the reptile during anaesthesia.
- **Environmental and Health Considerations:** Proper scavenging systems should be in place to manage environmental pollution and reduce health risks to the anaesthetist.
- **Aspects of Gaseous Anaesthesia Maintenance for Reptiles:** Inhalant gaseous anaesthesia is becoming the preferred method for anaesthetising reptiles during prolonged procedures due to its controllability and safety. Below are key aspects of this practice:

Behavioral changes

Snakes usually exhibit short period of agitation when first exposed to anesthetics, then gradually quieten until respiration becomes slow but regular. It is not easy to determined depth of anesthesia. The first indication snake can be safely removed from the container is loss of the righting reflex. Therefore, depth of anesthesia may be estimated by three signs.

1. Squeezing or pricking the tail is a particularly effective stimulant to snakes and loss of response indicates that surgical anesthesia has been produced.
2. When the tip of the tongue is grasped gently with forceps and withdrawn from its sheath, it remains extruded during surgical anesthesia, but at lighter level there is always marked resistance to with drawl.
3. During surgical anesthesia with all agents, the pupils are usually widely dilated.

4. As soon as snake is sufficiently sedated (losses of righting reflex) it is intubated. Rees modified T-piece which will allow artificial inflation of lung when required or IPPV set at low pressure and at a respiratory rate previously measured in conscious state. This is important since the expiratory movement of the single lung is too weak to expend gases through a conventional closed system. Fluid balance maintained by injecting isotonic saline to snakes of poor condition and if hemorrhage has occurred 20-24 ^{0}C temperature during maintenance.

3.6.4. Intubation

General Considerations

- **Ease of Intubation:** Reptiles generally lack an epiglottis, making intubation straightforward.
- **Glottis Position:** The glottis is relatively cranial in most species and remains closed at rest, opening during inspiration.
- **Saliva Production:** Reptiles produce little to no saliva at rest, reducing the risk of endotracheal tube blockage.

Species-Specific Intubation

Snakes

- Glottis Location: Rostral on the floor of the mouth, just caudal to the tongue sheath.
- Visibility: Easily visible when the mouth is opened.
- Procedure: Intubation can be performed in a conscious patient as reptiles lack a cough reflex. Use a wooden or plastic tongue depressor to open the mouth and insert the endotracheal tube during inspiration Fig. 7. Alternatively, use an induction agent prior to intubation.

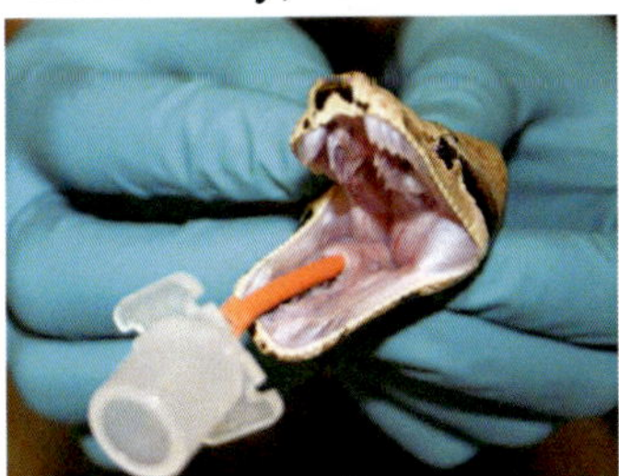

Intubation performed in a python

Chelonia (Turtles and Tortoises)

- Glottis Location: Slightly caudal at the base of the tongue.
- Trachea Length: Very short, requiring careful insertion of the endotracheal tube to avoid intubating only one bronchus.

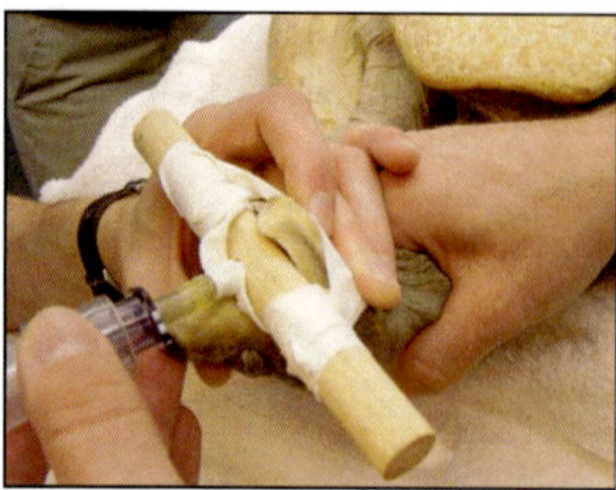

Intubation performed in Chelonia

- Induction Agents: Recommended due to their ability to breath-hold and the difficulty of extracting the head from the shell. Common agents include ketamine or propofol.

Lizards

- Glottis and Vocal Folds: Varies by species. Some geckos possess vocal folds.
- Intubation Method: Larger species may require a mouth gag to prevent biting down on the tube. Smaller species may not be suitable for intubation. Intubation can be performed consciously or after induction with an injectable or gaseous agent via face mask (Porter, 1972).

Crocodylia (Crocodiles and Alligators)

- Basihyal Fold: Acts as an epiglottis and must be depressed prior to intubation.
- Sedation Requirement: Due to their potential danger, these species require injectable chemical sedation or induction prior to intubation (Fleming 2014).

3.6.5. Intermittent Positive Pressure Ventilation (IPPV)

Introduction to IPPV in Reptiles

Intermittent positive pressure ventilation (IPPV) is a technique used to induce and maintain anaesthesia in reptiles, even in species known for breath-holding. It can be particularly advantageous because it can lead to rapid post-operative recovery by avoiding the use of injectable induction agents.

Key Points of IPPV in Reptiles

Inducing Anaesthesia

- Conscious Intubation: If intubation is performed on a conscious patient, anaesthesia can be induced using positive pressure ventilation.

- Breath-Holding Species: Positive pressure ventilation is effective even in species that typically hold their breath, inducing anaesthesia within 5–10 minutes.

Tidal Volume Variations: Tidal volume differs significantly among reptile species:

- Boas (Boiids): Approximately 12.5 mL/kg.
- Aquatic Turtles: Approximately 45 mL/kg.

Physiological Considerations

- Right-to-Left Pulmonary Shunting: This common condition in reptiles can be exacerbated by intracoelomic masses (e.g., eggs, food, effusions).
- Positioning: The position of the reptile during surgery can create a perfusion mismatch. For example, placing chelonians in dorsal recumbency for surgery can cause the liver and gastrointestinal tract to squash the dorsally located lungs.

Advantages of IPPV

- Rapid Recovery: Avoidance of injectable induction agents can lead to a quicker recovery post-operation.
- Anaesthesia Maintenance: It helps maintain anaesthesia and adequate oxygenation throughout the procedure.

IPPV Technique

- Ventilation Rate: Most reptiles require ventilation at a rate of six times per minute.
- Pressure: The ventilation pressure should be maintained at 10–15 cm of water.

Manual vs. Mechanical Ventilation

- Manual 'Bagging': With experience, manual ventilation can be performed by 'bagging' the patient with enough pressure to inflate the lungs without over-inflation.
- Mechanical Ventilation: Using a ventilator unit can simplify the process and ensure consistency.

Practical Tips

- Manual Ventilation: A rough guide for manual ventilation is to inflate the first two-fifths of the reptile's body at each cycle (Malley, 1997).

Additional Supportive Therapy

Recumbency

- **Chelonians:** Often placed in dorsal recumbency for intracoelomic surgery. Foam wedges or positional polystyrene-filled vacuum bags are essential to maintain stability.
- **Snakes:** May become extremely flaccid during surgery. They can be strapped to a long board or wedged in place with foam wedges or vacuum bags. Ensure the body wall is free of constraint and use IPPV if necessary for respiratory support.

Maintenance of Body Temperature

- Importance: Maintaining body temperature is crucial for successful recovery.
- Monitoring: Use a cloacal probe attached to a digital thermometer.
- Preferred Body Temperature (PBT): Maintain reptiles near their PBT, generally between 22–30°C.

Methods

- Heating Pads: Place the reptile on a circulating water or air heating pad during anaesthesia.
- Room Temperature: Keep the room temperature elevated to reduce heat loss.
- Warm Fluids: Administer warmed subcutaneous or intracoelomic fluids during and after surgery.
- Hot Water Bottles/Gloves: Can be used but must be wrapped in towelling to prevent direct contact. Monitor to ensure they don't cool down and draw heat away from the patient.
- Clear Drapes: Help retain heat.
- Light Sources: Many surgical lights radiate heat, aiding in temperature maintenance.

3.6.6. Monitoring Anaesthesia in Reptiles

Monitoring anaesthesia in reptiles requires adaptation of standard techniques due to their unique physiology and anatomy:

1. **Heart Rate and Rhythm**

   Challenges: Rough scales interfere with sound transmission, and the three-chambered heart reduces the clarity of the heartbeat.

Techniques

- Doppler Probe: Used over the heart outlet in lizards and Chelonia, and directly over the heart in snakes to monitor changes in rate and intensity of flow.
- ECG (Electrocardiography): Leads attached with alligator forceps on hypodermic needles can give an electrical trace of heart activity. Leads are placed cranial and caudal to the heart in snakes, and forelimb leads are placed cervically in some lizards.
- Heart Rate Formula: HR = 33.4 × [Body weight (kg)] $^{(-0.25)}$ assuming the reptile is at its preferred body temperature.

2. **Pulse Oximetry**
   - **Limitations:** Not useful in reptiles due to differences in hemoglobin structure.
3. **Capnography**
   - **Limitations:** Debateable due to cardiac shunting of blood in reptiles, which means that expired carbon dioxide levels do not reflect arterial levels.
4. **Respiratory Flow Monitors**
   - **Usefulness:** Limited due to the need for IPPV (Intermittent Positive Pressure Ventilation) in most reptile species.

   Other Considerations
   - Temperature Monitoring: Use a cloacal probe attached to a digital thermometer to monitor body temperature, aiming to maintain it near the preferred body temperature of the species.
   - Fluid Therapy: Essential for maintaining hydration and supporting recovery post-operatively.
   - Positioning: Ensure correct positioning during surgery to maintain stability and avoid complications.

3.6.7. Recovery and Analgesia in Reptiles

Recovery Management

Reptiles often recover rapidly from isoflurane anesthesia, but recovery may be prolonged if other injectable drugs like alphaxolone or ketamine are used. Here's how to manage recovery effectively:

- **Environment:** Keep the reptile calm, stress-free, and at its optimal preferred body temperature.

- **Intubation and IPPV:** Maintain intubation and continue intermittent positive pressure ventilation (IPPV) with oxygen until the reptile is breathing spontaneously.
- **Stimulating Respiration:** Use doxapram at a dose of 5 mg/kg by intramuscular or intravenous injection to help stimulate respiration. Remember, reptiles respond to falling blood partial pressure of oxygen (pO2) as a stimulus to breathe, not rising partial pressure of carbon dioxide (pCO2). IPPV with 100% oxygen may inhibit respiration; recovery using room air or increasing ventilation intervals to 20–30 seconds may be helpful.
- **Fluid Therapy:** Administer fluid therapy during recovery, especially for agents like ketamine cleared through the kidneys, to aid in speeding recovery.
- **Encouraging Eating:** Once fully recovered, encourage the reptile to eat. If anorectic, consider assist feeding or stomach tubing.

Signs of Recovery

- Snakes: May show sinuous movements from the tail end first.
- Chelonians: Will begin moving hindlimbs and then forelimbs.
- Lizards: Often exhibit uncoordinated hindlimb movements followed by forelimb movements.

Pain Assessment in Reptiles

Pain assessment in reptiles is challenging due to species variations. Consider the following indicators:

- General Signs of Pain: Immobility, anorexia, abnormal locomotion and posture, increased aggression, and changes in skin coloration (e.g., dull or dark).
- Species Variations: Assess behavior, environmental factors, locomotor activity, and miscellaneous attributes to gauge pain levels.

Analgesic Options

- **Buprenorphine:** Dose: 0.01–0.02 mg/kg IM; higher doses may be required in some species.
- **Morphine:** Dose: High doses effective in some species (e.g., 10 mg/kg in bearded dragons). Caution: High doses can depress respiration.
- **Butorphanol:** Effectiveness: Not always effective, but may have anesthetic-sparing effects.

- **NSAIDs:**
- **Carprofen:** 2–4 mg/kg IM initially, then 1–2 mg/kg every 24–72 hours.
- **Meloxicam:** 0.2 mg/kg orally every 24 hours; higher doses tolerated in some species (e.g., 5 mg/kg orally for 12 days in green iguanas). Caution: Potential nephrotoxic and gastrointestinal ulcerative side effects; monitor closely.
- **Tramadol:** Dose: 10–25 mg/kg orally; 11 mg/kg in bearded dragons; parenteral administration lasts 12–48 hours. Mechanism: Weak opioid activity on mu receptors.
- **Ketamine:** Use: Often used as adjunct to multimodal analgesia.Caution: Can cause bradycardia, hypotension, and reduced arterial pO2 levels.
- **Local Anesthesia:** Use: Ring blocks around amputations to reduce post-operative self-mutilation. Doses: Lidocaine up to 4 mg/kg (diluted at least 1:10); Bupivacaine up to 5 mg/kg.

The summary of the conclusions regarding anesthetic agents for reptiles are as follows:

- Older Anesthetic Agents: No longer viable for anesthesia and chemical restraint in reptiles.
- Narcotics: High doses required and legal constraints limit their value for reptiles.
- Ketamine: Useful alone or with tranquilizers like midazolam or diazepam, especially for breath-holding species. Readily available and inexpensive but may require large volumes and have prolonged recovery times.
- Tiletamine-Zolazepam: Effects are more variable than ketamine, but useful as an induction agent. Further research needed.
- Inhalation Agents: Most useful for surgical anesthesia. Rapid recovery and minimal metabolic side effects make them indispensable, especially for metabolically compromised patients. Isoflurane is preferred for debilitated reptiles due to elimination via lungs without requiring metabolism, despite its high cost.
- Agent Choice: Depends on the procedure's purpose. Some agents are better for short or nonpainful procedures, while others are suitable for general surgical anesthesia.
- Research: Continued research is essential to understand the effects of existing and new chemical restraint agents on reptiles.

3.7. Immobilization of Tortoises, Terrapins and Turtles

The order Chelonia, also sometimes referred to as Testudines, encompasses all turtles, tortoises, and terrapins. These fascinating reptiles are known for their unique shells, which offer them protection from predators.

The family Pleomedusidae includes black side necked turtle (*Pelusios subniger*) and helmeted turtle (*Pelomedusa subrata*).

The family Testudinidae includes box turtle (*Terrapen carolina*), painted turtle (*Chrysemys picta*), mobile terrapin (*Pseudemys elegans*) European pond turtle (*Emys orbicularis*), European pond turtle (*Geoemyda trijuga*), elephanline tortoise (*Geochelone elephantopus*) Red footed tortoise (*Geochelone carbonaria*), Horsefields tortoise (*Testudo horsfieldi*) Hermann's tortoise (*Testudo hermanni*) and African pancake tortoise (*Malacocherus tornieri*)

Here's a breakdown of the order Chelonia:

- **Ancient Lineage:** Chelonians are one of the oldest living reptile groups, with fossils dating back over 200 million years to the Triassic period. They predate even crocodiles and snakes!
- **Shell Shock:** Their most recognizable feature is their shell, which is actually made up of two parts. The upper part, called the carapace, is fused to their spine and ribs, while the lower part, known as the plastron, protects their belly.
- **Wide World Wanderers:** Chelonians can be found in a variety of habitats all over the world, including deserts, oceans, freshwater lakes and rivers, and even rainforests.

Here are some key characteristics of Chelonia:

- **Shell:** Their most defining feature is the hard shell, which is made up of two parts. The upper part, called the carapace, is fused to their spine and ribs. The lower part, called the plastron, protects their belly.
- **Beaks:** They don›t have teeth, but instead have a beak-like mouth used for tearing or grinding food.
- **Longevity:** They are known for their long lifespans. Some species, like the giant tortoise, can live for over 100 years.

While all chelonians share some common characteristics, there are also some key differences between turtles, tortoises, and terrapins:

Tortoises: These land-dwelling reptiles typically have high-domed shells and stumpy legs. They are herbivores and are known for their slow and methodical movements.

Tortoise

Turtles: These aquatic (water-dwelling) chelonians can be found in both freshwater and saltwater environments. They have streamlined bodies and webbed feet for swimming. Their diets can vary depending on the species, but they often include insects, fish, and aquatic plants.

Turtle

Terrapins: Similar to turtles, terrapins are semi-aquatic, meaning they spend time on land and in water. They tend to be smaller than turtles and have more colorful markings.

Terrapin

Anesthetic problems are posed by the anatomy and physiology of the chelonian respiratory system. The very low BMR varies with environmental temperature, and ability retract the head and limbs into the protective shell, and adaptation in some cases to aquatic or semi aquatic habits.

The lungs are well developed and, in some species, lie closely connected to the dorsal carapace. Ventilation results from changes in volume of the perivisceral cavity produced by inward movement of the limbs and skeletal girdles which is turn impose pressure changes on the lungs. Although muscular sheets close to the lungs have been described as diaphragm, they exercise relatively little effect compared with the mammalian diaphragm and exhalation probably results passively from the return of viscera to the resting position. The relatively large spaces within lungs suggest that concentration of O_2 and CO_2 equalize by diffusion. The implication for the anesthetic is that as in snakes, the expiratory pressure will be too weak to move gases within closed circuit, and that the ability to withhold hypoxia and survive on a single ventilator movement per hour can make induction with inhalational agents a frustrating experience (Green, 1982)

3.7.1. Recommended anesthetic techniques

It is not necessary to fast chelonians prior to surgery but they should be weighed and a clinical examination made for signs of ill health whenever practicable. Depending on species and the state of health, the effective body weight may only be 0.3 to 0.5 of the observed (body+shell). It is therefore safest to give injectable drugs at low computed dose levels and administer further increments as required. They can handle easily by hand. Some aggressive fresh water species handled by tranquilize by adding tricaine (1:1000) to their aquarium water prior to handling.

Since induction of anesthesia with inhalational agents is so unpredictable, it is suggested that parenteral administration of a suitable agent is the preferred method. Aquatic species should be kept damp throughout the anesthetic period.

Intravenous injection: Simpler approach is to make a small incision to expose jugular vein under narcosis and analgesia with etorphine (0.5 mg/kg) intramuscularly at gluteal muscle or ketamine (60 mg/kg) intramuscularly.

Intraperitoneal: Between neck and forelimb or by a caudal approach into the soft tissue between tail and hind limbs.

Estimation of depth of anesthesia is not easy. Respiration is best observed an alternating concavity and convexity in the soft tissue between hind limbs and tail. If narcosis becomes too deep, the limb muscle controlling ventilatory movements may be paralyzed and artificial ventilation may than be necessary. The eyes give no indication of depth of anesthesia. Corneal blink reflex is often present even during deep anesthesia. Loss of muscle tone in the head, neck and limbs provides the best indication of narcosis and withdrawal of limbs to pin prick is a good guide to analgesia.

Ketamine: Dose 60 mg/kg, intramuscular in gluteal muscle. The induction times is 30 min and have 60 min duration of anesthesia. The recovery time is 24 hours. Anesthesia can be depended with N_2O: O_2 (1:1) and 3% halothane or 2% methoxyflurane supplied by mask insufflation catheter.

Pentobarbitone: Dose is 10 mg/kg intraperitoneally. It should be administrated 30 min after injection chlorpromazine give at the rate of 10 mg/kg intramuscularly. Induction time is 15-30 minutes and recovery occur in 6 to12 hours. If pentobarbitone used alone then the dose is 16-18 mg/kg given intraperitoneally. Induction time is 60-90 min with adequate muscle relaxation. Recovery occurs in 3-4 hours.

Following anesthesia, the terrestrial species should be allowed to recover in an ambient temperature of 18-20 ^{0}C and high humidity (50-60%). Aquatic forms should be maintained at 16-18 ^{0}C with the body wrapped in dump towels.

3.8. Immobilization of Lizard, Alligators and Crocodiles

3.8.1. Lizards

Lizards are the most diverse group of reptiles within the order Squamata. They come in a wide variety of shapes, sizes, and colors, and can be found on every continent except Antarctica. The Family Laceritidae includes Wall lizard (*Lacerta muralis*), Sand lizard (*Lacerta agilis*), Emerala lizard (*Lacerta viridis*) and Ruin lizard (*Lacerta sicula*).

Some well-known lizard species include:

- **Geckos:** These nocturnal lizards have soft skin and adhesive toe pads that allow them to climb walls.

Gecko Lizard

- **Iguanas:** These large, herbivorous lizards are found in tropical and subtropical regions of the Americas. They belong to Family Iguanidae. This includes Common Iguana, American anole (*Anolis carolinesis*), Rhinoceros iguana (*Cyclura cornuta*), Ricordis iguana (*Cyclura ricordi*)

Iguana Lizard

- **Chameleons:** These arboreal lizards are known for their ability to change color and their independently moving eyes. The Family Chamaelontidae includes Two flapped chameleon (Chameleo dilepis).

Chameleon Lizard

- **Skinks:** These slender-bodied lizards have smooth, shiny scales and can lose their tails as a defence mechanism.

Skink Lizard

- **Komodo Dragons:** These giant lizards are the largest living species of lizard, growing up to 10 feet in length. The Family Varanidae includes Komodo dragon (*Varanus komodoensis*), Water monitor (*Varanus salvator*), Nile monitor (*Varanus niloticus*) and Bengal monitor (*Varanus bengalensis*).

Komodo Dragon Lizard

Some key characteristics of lizards include:

- Scaly skin that they shed periodically
- Elongated body with four legs (some legless lizards exist)
- Claws on their feet
- External ear openings (except for some legless lizards)
- Varied eyelids, some with movable eyelids and others with a single transparent membrane covering the eye
- Elongated tongues (in some species) used for tasting and capturing prey
- Varied diets (omnivores, insectivores, herbivores)

Lizards come in a wide range of shapes and sizes, from tiny geckos that can fit on your fingertip to large iguanas that can grow several feet long. They also exhibit a diverse range of diets, habitats, and behaviours.

Inhalational agents

For lizards 5-10% of halothane or 3% methoxyflurane in a jar or in a box soaked with cotton gauze. Induction time is 10 min using halothane and 20-30 min for methoxyflurane. Recovery time is 20 and 60 min, respectively. Maintenance is best achieved by N_2O: O_2 (1:1) (1:1) 1:1 halothane or methoxyflurane delivered through a nose cone, mask or oral catheter.

Recovery and post-surgical therapy

Low temperatures will depress metabolism and hinder recovery while temperatures accelerate metabolism but may produce total hypoxia so ideal temperature 22-24^{0}C should be provided.

3.8.2. Alligators and Crocodiles

Crocodiles belong to a completely different order, Crocodilia. They are not closely related to lizards, although both are reptiles. They are more closely related to birds, which are also descended from the archosaur group. The family Crocodylidae includes saltwater crocodile (*Crocodylus porosus*) and American crocodile (*Crocodylus acutus*). The family Alligatoridae includes American alligator (*Alligator mississippiensis*).

Order Crocodilia (Crocodile)

Here are some key characteristics of Crocodilia:

- Thick, leathery hide with bony plates embedded for protection
- Powerful jaws and sharp teeth for catching prey
- Strong legs and webbed feet for swimming
- Nostrils and eyes located on the top of their head, allowing them to breathe and see while most of their body is submerged
- Salt glands on their tongue to help regulate salt intake

Crocodilians are large, semi-aquatic predators found in freshwater and saltwater habitats throughout the tropics and are distantly related to lizards. They are apex predators in their ecosystems and have remained relatively unchanged for millions of years. There are three living families of crocodilians: crocodiles, alligators, and gavials.

- **Crocodiles:** These are the most widespread type of crocodilian, found in Africa, Asia, the Americas, and Australia.

Crocodile

- **Alligators:** These are found only in freshwater habitats in the Americas and are closely related to crocodiles.

Alligator

- **Caimans:** These are another group of New World crocodilians, found in South America.

Caiman

- **Gavials:** These are easily recognizable by their long, slender snouts, which are an adaptation for fishing.

Gavial

Key characteristics of crocodilians include

- Large, elongated body with short legs and a powerful tail
- Thick, hide-like skin with bony plates for protection

- Long snouts with sharp teeth
- Strong jaws and powerful bite
- Aquatic lifestyle, with nostrils and eyes positioned high on the head for stealthy hunting
- While both lizards and crocodilians are reptiles, they belong to distinct evolutionary lineages and have significant differences in anatomy, behaviour, and habitat.

These groups have ectothermic characters of reptiles. Crocodiles and alligators detoxify drugs very slowly and as a result, recovery from anesthesia or chemical immobilization is always prolonged.

Recommended anesthetic techniques

Anesthesia provably best carried out at an ambient temperature of 22-24 ^{0}C. Fasting prior to anesthesia is not necessary but the animal should be carefully observed for signs of disease or inanition. Reptiles in poor health should be given a quarter of the estimated normal dose. The depth of anesthesia attained is best judged by loss of the righting reflex, loss of sensation assessed by withdrawal on pinching or pricking the feet, and by the frequency of respiration.

Injectable techniques

Diazepam followed by succinyl choline chloride was administrated intramuscularly to alligator at the mean dosage of 0.37 mg/kg and 0.24 mg/kg, respectively. This drug combination reduced stress and allowed adequate immobilization for restraint and handling this combination a good alternative for immobilization alligators (Spiegel, 1984).

Ketamine: It is the drug of choice. The initial dose is 10-25 mg/ kg given at the base of the tail. The induction time is 50 min. Duration of anesthesia is 4-7 hours in crocodiles and alligators, 2-6 hours in lizards.

Further increment by using N_2O: O_2 (1:1) and 1-2% of halothane or 3% methoxyflurane

Tricaine: The dose is 80 mg/kg intramuscular in alligators and crocodiles. The induction time is 10 minutes and recovery time is 10 hours

4

Chemical Restraint of Aves

4.1. Classification

The class Aves, also known as birds, is a group of warm-blooded vertebrates characterized by feathers, toothless beaked jaws, laying hard-shelled eggs, a high metabolic rate, a four-chambered heart, and a strong yet lightweight skeleton. Birds are found worldwide and come in a variety of sizes, from the tiny humming bird to the massive ostrich.

The classification of Aves can be broken down into two major subgroups: Palaeognathae and Neognathae.

4.1.1. Palaeognathae: Also known as ratites, these are flightless birds with flat sterna (breastbones) that lack keels, which are important for flight muscle attachment. Examples of palaeognathae include ostriches, emus, rheas, and kiwis.

4.1.2. Neognathae: This group includes all other birds, both flying and flightless. They have keeled sterna and a more complex palate structure than paleognathae. Neognathae are further divided into many different orders, such as:

a. **Falconiformes:** birds of prey, such as hawks, eagles, falcons, and vultures. It includes Andean condor (*Vultur gryphus*), King vultures (*Sarcoramphus papa*), Black vultures (*Coragyps atratus*), Turkey vutures (*Cathartes aura*), California condor (*Gymnogyps californianus*), Swallow-tailed kite (*Elanoides forficatus*), Northern goshawk (*Accipiter genitilis*), Common buzzard (*Buteo buteo*), Rough- legged buzzard or rough-legged hawk (*Buteo lagopus*), Red shouldered hawk (*Buteo lineatus*), Broad winged hawk (*Buteo platypterus*), Harpy eagle (*Harpia harpyja*), Golden eagle (*Aquila chrysaetos*), Tawny eagle (*Aquila rapax*), Bald eagle (*Haliaeetus leucocephalus*), Griffon vultures (*Gyps fulvus*), Egyptian vultures (*Neophron perenopterus*), Osprey or sea hawk (*Pandian haliaetus*), Gyrfalcon (*Falco rusticolus*), and Merlian or pigeon hawk (*Falco columbarius*).

b. **Passeriformes:** perching birds, the largest and most diverse order of birds, including songbirds, crows, jays, and sparrows. It includes

Canary (*Serinus canaries*), Zebra finch (*Taeniopygia castanotis*), Jaw sparrow (*Paida oryzivasa*), Mynah birds (*Gracula religiosa*), Bengalese finch (*Lonchura domestica*), Budgerigar or parakeet (*Melopsittaacus undulates*), and Cockatiel (*Nymphicus hollandicus*).

c. **Anseriformes:** Waterfowl, such as ducks, geese, and swans. These includes Mute swan (*Cygnus olor*), Black swan (*Cygnus atratus*), Black-necked swan (*Cygnus melanocoryphus*), Trumpeter swan (*Cygnus cygnus*), Whistling swan (*Cygnus columbianus*), Coscoroba swan (*Coscoroba coscoroba*), Canada goose (*Branta Canadensis*), Barnacte goose (*Branta leucopsis*), Snow goose (*Anser caerulescens*), Emperor goose (*Anser canagicus*), Hawiian goose (*Branta sandvicensis*), Bar-headed goose (*Anser indidcus*), Greylag goose (*Anser anser*), White-fronted goose (*Anser albiforons*), Cape Barren goose (*Cereopsis novaehollandiae*), Andean goose (*Chloephaga melanoptera*), Egyptian goose (*Alopochen aegyptiaca*), Mallards (*Anas platyrhynchos*), Common shel duck (*Tadornatadema*), Muscory duck (*Cairina moschata*), Ringed teal (C*allonetta leucophrys*), Wood duck (*Aix sponsa*), Mandarian duck (*Aix galericulata*), American wigeon (*Mareca americana*), American black duck (*Anas rubripes*), Blue-winged teal (*Spatula discors*), Cinnamon teal (*Spatula cyanoptera*), and Common eider or Cuddy's duck (*Somateria mollissima*).

d. **Columbiformes:** Pigeons and doves. It includes Rock pigeon, rock dove, or common pigeon (*Columba livia*), Mourning dove (*Zenaida macroura*), and Long tailed passenger pigeon (*Ectopistes migratorius*).

e. **Strigiformes:** Owls. It includes barn owl (*Tyto alba*), Screech owl (*Megascops asio*), Great horned owl or tiger owl (*Bubo virginianus*), Europian eagle owls (*Bubo bubo*), Snowy owl or white owl (*Bubo scandiacus*), Burrowing owls (*Athene cunicularia*), and barred owl (*strix varia*).

f. **Charadriiformes:** Shorebirds and seabirds, such as gulls, terns, sandpipers, and loons.

g. **Psittaciformes:** Parrots. Lovebird is the common name for the genus *Agapornis*, a small group of parrots in the Old World parrot family Psittaculidae. Of the nine species in the genus, all are native to the African continent, with the grey-headed lovebird being native to the African island of Madagascar. These birds excellent mimic, and are easily taught to talk. It includes Peach faced love bird (*Agapornis roseicollis*), Black masked love bird (*A. personata*), Yellow headed Amazon parrot (*Amazona ochrocephala*), Atrican gray parrot (*Psittacus erithacus*),

Macaws – (*Hyacinthine macaw*) (*Anorhynchus hyacinthinus*), Blue and gold macaw (*Ara ararauna*), and Scarlet macaw (*Ara macao*).

h. **Sphenisciformes:** Penguins. It includes Humboldt penguin (*Spheniscus humboldti*), King penguin (*Aptenodytes patagonicus*), Rockhopper penguin (*Eudyptes chrysocome*), Emperor penguin (*Aptenodytes forsteri*), and Fairy penguin or blue penguin (*Eudyptula minor*)

This is just a small sampling of the many bird orders within Neognathae. With over 10,000 living species, the class Aves is incredibly diverse and plays a vital role in ecosystems around the world.

4.2. Restraint of bird

Restraint of birds, also known as avian restraint, is the act of holding a bird in a way that minimizes stress and the risk of injury to both the bird and the handler. It is a necessary procedure for veterinarians to perform examinations and administer medications, and it can also be necessary for bird banders and researchers.It's important to note that restraint should only be performed by trained professionals or experienced bird handlers, as improper restraint can cause serious injuries or stress to the bird.There are a number of different methods for restraining birds, depending on the size and temperament of the bird, as well as the purpose of the restraint. Here are some common methods:

4.2.1. Towel wrap: This is a common method for restraining small to medium-sized birds. A soft towel is wrapped around the bird's body, covering the wings and legs. The bird's head may be left exposed, or it may be gently restrained under the towel.

Towel wrap bird restraint

4.2.2. Physical restraint: This method involves holding the bird in the handler's hands. The bird's body should be held firmly but gently, with the wings and legs restrained. The handler should be careful not to compress the bird's chest, as this can make it difficult for the bird to breathe.

Physical bird restraint

4.2.3. Chemical restraint: In some cases, it may be necessary to chemically restrain a bird before handling it. This is typically only done by a veterinarian.

There are different techniques for restraining birds, depending on the size and temperament of the bird. Here are some general guidelines:

- ***Small birds:*** Small birds can be gently cupped in one hand with the head secured between the thumb and index finger.
- ***Medium-sized birds:*** Medium-sized birds can be restrained using a towel. The bird is gently wrapped in the towel, with the head and legs secured.
- ***Large birds:*** Large birds may require two people to restrain. One person can hold the bird's body, while the other person secures the head and legs.

Here are some important things to keep in mind when restraining a bird:

- ***Minimize stress:*** The goal is to restrain the bird as quickly and safely as possible to minimize stress.
- ***Maintain respiration:*** Do not compress the bird's chest, as this can make it difficult for the bird to breathe.
- ***Avoid injury:*** Be careful not to injure the bird's wings, legs, or beak.
- ***Use proper equipment:*** Some birds may require specialized equipment for restraint, such as cones or hoods.

It is important to note that restraint can be stressful for birds, so it should only be done when necessary. When restraining a bird, it is important to handle it calmly and gently. The bird should be kept in a quiet, well-ventilated area, and the restraint procedure should be kept as short as possible.Here are some additional tips for restraining birds:

- Wear gloves to protect yourself from bites and scratches.
- Have a helper on hand to assist you, if necessary.

- Be familiar with the bird's anatomy and physiology.
- Observe the bird's behaviour for signs of stress, such as panting, fluffed feathers, or aggression.
- If the bird becomes too stressed, stop the restraint procedure and try again later.
- Birds are very sensitive to hypothermia, so it is important to maintain their body temperature during anaesthesia.
- Because birds have a very efficient respiratory system, they can quickly become overdosed on anaesthetic agents. Careful monitoring is essential to ensuring the safety of the bird.

4.3. Anesthetic technique

In avian anaesthesia, the main concern is providing adequate anaesthesia while minimizing the risk of complications. Unlike mammals, birds have a unique respiratory system that requires special considerations when administering anaesthesia.Inhalant anaesthesia is the most common technique used for Aves (birds). This involves delivering a gaseous anaesthetic agent, usually isoflurane or sevoflurane, mixed with oxygen through a mask or endotracheal tube.Here's a general overview of the process:

1. ***Pre-anaesthetic assessment:*** A thorough examination is performed to assess the bird's overall health and identify any potential risks.
2. ***Induction:*** The bird is placed in an induction chamber with a high concentration of isoflurane or sevoflurane to induce anaesthesia.
3. ***Maintenance:*** Once anesthetized, the bird is maintained on a lower concentration of the inhalant anaesthetic agent mixed with oxygen.
4. ***Monitoring:*** Vital signs such as heart rate, respiratory rate, and oxygen saturation are closely monitored throughout the procedure.
5. ***Recovery:*** The bird is gradually weaned off the anaesthetic agent and allowed to recover in a warm, oxygen-enriched environment.

Other anaesthetic techniques can be used in conjunction with inhalant anaesthesia, such as:

- **Injection anaesthesia:** This may be used for short procedures or in birds that are difficult to mask.
- **Regional anaesthesia:** This involves blocking nerves in a specific area of the body.

  The choice of anaesthetic technique will depend on the type of procedure being performed, the size and species of the bird, and its overall health status.

Several aspects of a bird's behavior, anatomy, and physiology are relevant in the selection of an anesthetic method. These include the lower degree of pain apparently experienced during surgical interference as compared with mammals, the very high basal metabolic rate, particularly in small species. Thermoregulation is both less efficient and adaptable in birds than in mammals, and the potential for heat is further increased during anesthesia. In young chicks' prevention of heat loss is therefore particularly important in birds, especially if they are small or young. Evaporation is lost from the respiratory tract if the respiratory rate increases and when anesthetic gases are supplied at high flow rates. Plumage should remain intact if possible, and the minimum area possible should be prepared for surgery with water- or alcohol-based solutions. There is an increased risk of hypoglycemia from food deprivation. The rapid heart rate changes drastically in response to tears or physical exertion. Birds should not be maintained on their backs for long periods since hypotension resulting from decreased venous return is likely to develop.

Acute cardiovascular disease may result from restraining, securing the wings to the back, and fastening the legs together with adhesive tape. The respiratory system is adapted for respiration during level flight, soaring, and diving at speed, and this has particular implications for inhalational anesthesia. Volatile anesthetics have proved satisfactory but must be given with caution because of the air sac system, which may allow a dangerous over concertation of the anesthetic before the anesthesiologist realizes this has occurred. Poorly ventilated air sacs may be a source of high concentrations of inhalation anesthetic agents that may subsequently be released into the circulation post-anesthetic administration of oxygen for five minutes or longer, which is therefore important.

1. The dose of ketamine per kg body weight for birds is inversely proportional to the weight of the bird.
2. Heavier doses of ketamine will produce a deeper plane of anaesthesia. Hence, a low dosage may only achieve restraint, while a larger dose will produce anesthesia suitable for diagnosis and minor surgical procedures.

Boever and Wright (1975) have given base-line figures for ketamine dosages.

1. Birds weighing less than 100 g (canaries, tenches, parakeets, and small birds) 0.5-1 mg/kg body weight.
2. Birds weighing between 250 and 500 g (parrots, pigeons, and other medium – sized birds) 0.05-0.1 mg/kg body weight.
3. Birds weighing between 500 and 3000 g (chickens, owls, hawks, and other larger birds) 0.02-0.1 mg/kg body weight.

4. Birds weighing more than 3000 g (ducks, swans, and larger birds) 0.02-0.05 mg/kg body weight.

Restraint and immobilization are defined as a moderate central nervous system depression in which the bird is calm and free; there is little or no movement unless the bird is provoked by painful stimuli; corneal and pedal reflexes are present. Anesthesia is described as a loss of sensation; there is no resistance and pedal, corneal reflexes are absent. The critical time is the recovery period during which the bird is excited, thrashing, and lacking co-ordination. Frenzelled wing flapping and head shaking are common, and care must be taken, especially with long-necked birds such as ducks and swans. To provide a padded recovery area, decreased body temperature is due to lowered muscle tone; a warm recovery area is essential.

Ketamine HCl (2 mg) is sufficient for performing general surgery on parakeets (Mandeker, 1972). If a longer period of surgical anesthesia is required, it can be induced with ketamine HCl (1 mg) maintained with methoxyflurane induction of anesthesia within 3 minutes. The recovery from anesthesia occurs in 20-30 minutes.

Methoxyflurane is probably the safest volatile agent for use in birds. Small birds may be anesthetized by placing them in a closed container. One-quart instant coffee jar into which 0.1 ml of methoxyflurane is injected. The induction requires 30 to 45 seconds, following which the bird may be removed from the jar and anesthesia maintained with a methoxyflurane- soaked cotton held at the birds nostrils mask for small birds can also be prepared from a disposable syringe container in which cotton is inserted; for larger birds, a larger jar can be used afterwards, incubated connected to an anesthetic machine (Green, 1982).

Ether can be used for brief anesthesia. The bird is held in a recumbent position, and ether is sprayed from a 2 ml syringe equipped with a 26-gauge, one-quarter-inch needle into a nostril from a distance of about 1 inch. Administration must be discontinued as soon as struggling ceases and should be resumed at intervals as the bird begins to regain consciousness. Ethyl chloride is also used in the same fashion.

Intramuscular pentobarbital sodium can also be used. Diluted 1 ml is added to 8 ml of sterile water. For smaller birds' tuberculin syringe should be used. Oral pentobarbital is safer for parakeets and canaries than that given parentally. The birds should be fasted for 24 hours Two mg of pentobarbital sodium diluted in 0.24 ml of water and given orally. If necessary second dose may be given after 15 minutes. Procaine is toxic for parakeets and should never be used in this species. Procaine penicillin is also contra indicated (Green, 1982).

Mandelker (1971) described simple devices to simplify the anaesthesia of birds with methoxyflurane (metofane). One such aid is a large, transparent plastic laundry bag placed over the cage and sealed. A hole is cut into the bag, where it is to be attached to the anesthetic machine. High O_2 inflow (1-5 L/min) is used, and the vaporizer setting is adjusted to the fully open position. In most cases, a light plane of anesthesia permitted the removal of the bird from the cage for quick non-surgical procedures. The use of the bag over the cage permits the operator to evaluate the progress of anesthetization and eliminates unnecessary handling of the birds before they are anesthetized. If prolonged surgical anesthesia is required, a small plastic sandwich bag may be used as a face mask. A hole is cut into the closed end of the bag, and a rubber ring, made from the base of an infant feeding nipple, is glued to this hole. The holes in the bag and ring should be just large enough to fit over the bird's head. The open end of the bag is then tied to the body of the anesthesia unit. This method permits surgical access to the entire body of the bird below the neck. In most instances, a parakeet maintained an anesthetic reading of 4 -6 well over 1 hour. For parakeets and other large birds, the covered cage technique can be used to initiate anesthesia; however, a small nose cone is generally used to administer methoxyflurane and O_2 to large birds. After induction of anesthesia with the nose cone, a medium-sized plastic gag is placed over the bird's, head and neck, gently around the neck. A hole is cut into the closed end of the bag for attachment to the anesthesia machine. A warm air blower directed at the bird often helps hasten recovery from the anesthetic after a long surgical operation. The author used this method successfully with parakeets, canaries, parrots, macaws, and goose.

4.4. Capture of feral birds

Tribromoethanol is used in capturing wild turkeys. Dose: Undiluted 0.75 ml orally with corn, capturing a wild mourning dove using a methohexital mixture of wheat and cracked corn. Narcosis occurs within 30 minutes. Deep narcosis occurs about two hours after feeding. Tribromoethanol was the only compound to satisfy the criteria of initial tests for anesthesia in mallard doves (David, 1972). The mean duration of the induction, immobilization, and recovery periods was 2.4 minutes, 8.7 minutes, and 1.3 hours, respectively, at the medium effective dose for immobilization of 100 mg/kg body weight. The median lethal dosage (LD50) is 400 mg/kg.

Severe limitations and hazards are inherent in the technique of bird capture with treated baits (David, 1972).

1. Lack of control over the amount of treated bait consumed by individuals.
2. There is variability in the responses of target subjects because of differences in sex, age, health, and stress.
3. Paucity of information as to the potential effects of drugs on behavior, physiologic processes, and survival.
4. Escape of partially narcotized individuals from the bait side.
5. Potential vulnerability of non-target species and
6. There is a possibility that chemical residue might be deposited in the edible tissues of treated birds. For this reason, a need exists for an intensive and thorough evaluation of the effectiveness, safety, and practicability of this method of bird capture before adopting it for general use.

4.5 Anesthesia in Raptors

For raptors (also known as birds of prey, a hyper carnivorous bird species that actively hunts and feeds on other vertebrates), the recommended technique is intravenous ketamine and diazepam (Redig and Duck, 1976). The dosage is 30-40 mg / kg body weight of ketamine HCl and 1-15 mg/kg body weight of diazepam. Intravenous injections of ketamine HCl and diazepam yielded better results with both lower and narrower dosages in owls and raptors. Owls had considerably more species-specific variation than did hawks. Tolazoline was useful in controlling the duration of xylazine HCl- and ketamine HCl induced anesthesia in turkey vultures (Allen, 1986).

Breast muscles are so important in flying, so falcons should never be injected into the breast muscles. Instead of breast muscles, leg muscles can be used. The femoral vein and jugular vein are readily accessible in falcons for intravenous use. Reserpine in falcons 1-2 mg/lb orally in meat bird progressively relaxed shares of hypnotic state develop temperature drops down 108 ^{0}F to 100 ^{0}F and tranquilization lasts 5-6 days.

Intravenous ketamine HCl (35 mg/kg) and diazepam (1-1.5 mg/kg) have been used successfully for 4-5 minutes of anesthesia. Anesthesia was induced by introducing 3% isoflurane in O_2 (3 L/min) and N_2O (6 L/min) into the transporters (a fiberglass transporter tightly enclosed in plastic with a small slot for observation). The bird was laterally recumbent for approximately 5 minutes and was removed from the plastic enclosure. O_2 and N_2O were continued at 3 and 6 L/min, respectively, and the isoflurane concentration was reduced to 2% (Richard *et al.*, 1985). After two minutes of anesthetic mask induction, the condor had relaxed sufficiently to allow the insertion of a cuffed endotracheal tube (60 mm inside diameter, 1 mm outside diameter). The cuff was not inflated to avoid injury or rupture of the tracheal rings. The O_2 and N_2O flow rates were reduced to 1 and 2 L/min, respectively. The vaporizer setting was adjusted to between 1.0 + and 2.5 L throughout surgery to maintain a slightly positive foe pinch reflex 15 minutes before surgery. N_2O was discontinued. O_2 flow increased to 3 L/min, and the vaporizer setting was decreased to 0.5%. After completing surgery, the vaporizer turned off. O_2 continued for 5 minutes.

4.6 Anesthesia in Penguins

Anesthetizing penguins follows similar principles to avian anaesthesia in general, but with some specific considerations for their unique physiology.

- **Inhalant anaesthesia:** Similar to other birds, inhalant agents like isoflurane or sevoflurane with oxygen are most common. Induction can be via mask or endotracheal tube (after considering tracheal septum placement).
- **Injectable anaesthesia:** Combinations of drugs like medetomidine (a sedative), ketamine (pain relief), and butorphanol (analgesic) can be used for short procedures or as a pre-medication before inhalants.

Challenges and Considerations

- **Tracheal Septum:** Penguins have a unique structure in their windpipe (trachea) called a tracheal septum. This partition needs to be located during endotracheal intubation for proper ventilation.
- **Temperature regulation:** Penguins are susceptible to both hypothermia (low body temperature) and hyperthermia (high body temperature) during anaesthesia. Maintaining their body temperature is crucial.
- **Respiratory system:** Careful monitoring of respiration is essential due to their efficient respiratory system, making them prone to overdoses on anaesthetic agents.

The administration of 2.5% thiamylal intravenously in a penguin has been reported. An injection was made into a vein on the posteromedial aspect of one flipper. Additional increments were given as needed to allow the removal of tumors, etc.

5

Chemical Restraint of Rodents and Lagomorphs

Rodents and lagomorphs are two distinct groups of mammals, often mistaken for each other due to some shared physical characteristics. However, they belong to different taxonomic orders and possess several key differences

5.1. Rodents

They belong to the order Rodentia and the example species include rats, mice, squirrels, hamsters, guinea pigs, beavers, and porcupines. Order Rodentia is the largest and most diversified mammalian group, differing widely in size, behavior, feeding habits, anatomy, and physiology. Based on the anatomy and functional differences of masseter muscles, it is divided into 3 suborders viz., Caviomorpha, Myomorpha, and Sciuromorpha. Caviomorpha is also called hystrichomorpha and includes guinea pigs, chinchillas and degu. The suborder Myomorpha includes mice, rats, hamsters and gerbils. These animals have the strongest ability to gnaw among animals in all 3 suborders. Sciuridae is a family that includes small or medium size rodents. This family consists of tree squirrels, ground or medium size squirrels, and large size squirrels also known as marmots or prairie dogs (*Cynomys ludovicianus)*. Smaller and less bushy-tailed squirrels are known as chipmunks (*Tamias striatus)* and flying squirrels (*Glaucomys volans)*.

Rodents have two pairs of continuously growing incisors in both upper and lower jaws (total of four). The diet is primarily herbivorous, but some rodents are omnivorous.

The distinguishing features are:

- Presence of a single pair of prominent, chisel-like incisors in both upper and lower jaws.
- Large, ever-growing incisors used for gnawing and feeding.
- Diastema: a gap between the incisors and the cheek teeth.
- Rodents have continually grown and erupting teeth without an anatomic root in the maxilla and mandible and are known as elodont aradicular teeth.

- Rodents have a single set of teeth known as monophyodont and 1 pair of incisors called simplicidentate.

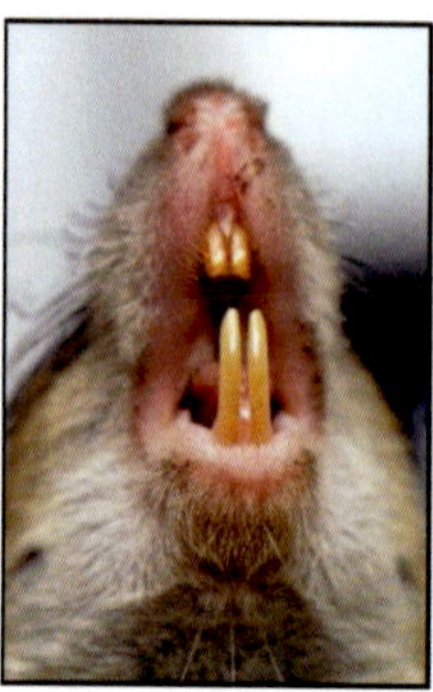

Field rat incisors

5.2. Lagomorphs

Lagomorphs belong to Order Lagomorpha of which there are two living families: the Leporidae (rabbits and hares) and the Ochotonidae (pikas). Lagomorphs are a group of mammals distributed throughout the world known for their large ears and amazing speed. Although they resemble rodents in some details, they also differ in many ways. Lagomorphs belong to a different family with their own characteristics. There used to be more than 200 species, but today we know 80 living species of lagomorphs belonging to 2 families with 13 genera. Rabbits are elodont and hypsodont (high crowned).

5.3. Differences between families of lagomorphs

Rabbits and hares move by jumping, pushing off with their strong hind legs and using their forelimbs to soften the impact on landing. Pikas lack certain skeletal modifications present in leporids, such as a highly arched skull, an upright posture of the head, strong hind limbs and pelvic girdle, and long limbs. Also, pikas have a short nasal region and entirely lack a supraorbital foramen, while leporids have prominent supraorbital foramina and nasal regions.

The anatomy of lagomorphs is specifically designed to flee from predators. Their long ears, the position of their eyes, and their powerful limbs are structures perfectly designed to detect danger and escape quickly. Additionally, many of the taxonomic features important in diagnosing lagomorphs are related to their herbivorous habits. Let us take a look at some of their most important physical characteristics:

Lagomorphs have one pair of prominent incisors in the upper jaw and two pairs (one large and one small) in the lower jaw (total of four). The distinguishing features are:

- Presence of two pairs of incisors in the upper jaw (one large and one small pair behind)
- Smaller peg-like second pair of incisors located behind the larger incisors in the upper jaw
- Cheek teeth adapted for grinding plant material
- Like rodents, their teeth grow permanently.
- The diet is strictly herbivorous.
- Their body is covered with fur that varies in color.
- They are placental mammals and have mammary glands.
- They are endothermic, which means their body heat does not depend on the external environment.
- The projection on the frontal bone near the upper posterior edge of the skull is slightly fused, a feature that is different in other mammals.
- They have large ears, although some more than others, and a keen sense of hearing.
- Their eyes are arranged on the top of the head and are located on each side of the head, allowing them to see the environment almost 360 degrees, which enables them to detect potential threats.
- They have four limbs characterized by strong structures that give them agility and speed when walking.
- The weight can vary depending on the species, but generally it does not exceed 5 kg; there are even species that weigh little more than 100 grams.
- They have an excellent sense of smell.
- To digest plant food, they have a well-developed digestive system.
- Some of the feces they produce are called "cecotrophs" and they ingest it as soon as they defecate it, as it still contains nutrients that can be utilized by the animal. For this reason, they can often be seen eating their feces.
- Females are larger than males, a characteristic of lagomorphs that is rare in other mammals.
- Their lower legs are covered with hair and have no pads.
- Their eyes are large, and their night vision is good, which is due to their predominantly nocturnal or crepuscular lifestyle.

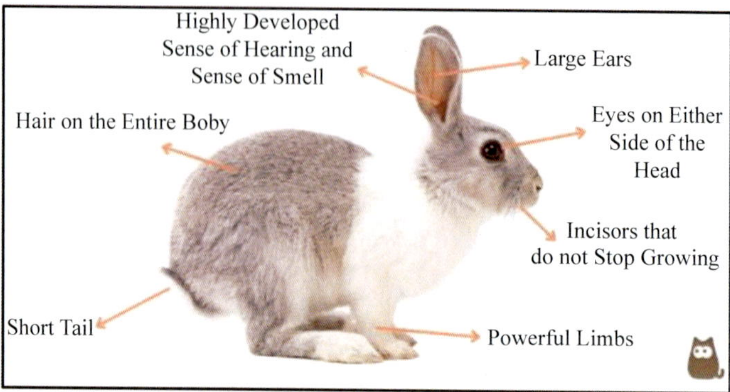

Characteristics of Lagomorphs

Example species of Lagomorphs are rabbits, hares, pikas (Roch rabbits), pikas or rock rabbit *(Ochotona princeps),* cotton tail rabbit (*Sylvilagus flovidanus)* and hares (Le pus spp).

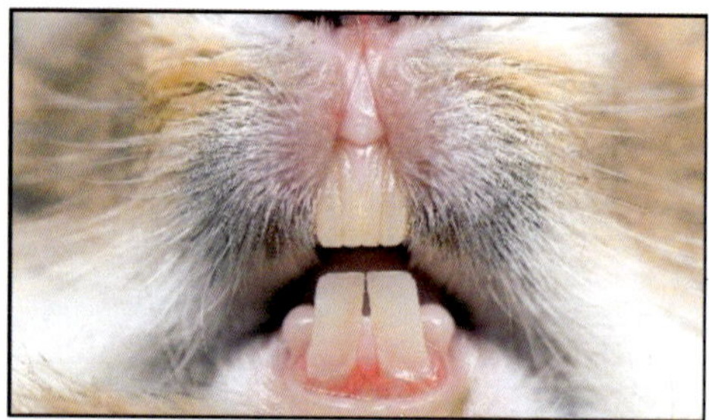

Rabbit teeth

5.4. Key Differences between rodents and lagomorphs

The most significant difference between rodents and lagomorphs lies in their dentition. While both groups have continuously growing incisors, rodents have only one pair in each jaw, while lagomorphs have two pairs in the upper jaw (one large and one small behind). Additionally, lagomorphs possess more complex cheek teeth compared to rodents.

Other notable distinctions between rodents and lagomorphs

- ***Diet:*** While most rodents are herbivores, some are omnivorous, occasionally consuming insects or other animal matter. In contrast, lagomorphs are strictly herbivorous.
- ***Habitat:*** Rodents exhibit a wider range of habitat preferences, including forests, grasslands, deserts, and even aquatic environments. Lagomorphs are primarily terrestrial and prefer open habitats like meadows, grasslands, and rocky areas.

- ***Behaviour:*** Many rodents are social creatures, living in colonies or groups. Lagomorphs tend to be more solitary, except during breeding season.
- Os penis is present in rodents but it is absent in lagomorph
- Scrotum is posterior to penis in rodents but it is anterior to penis in lagomorphs

Understanding these differences is crucial for proper identification, care, and management of these diverse and ecologically important groups of mammals.

5.5. Restraint of Rodents and Lagomorphs

Chemical restraint is a crucial procedure used in veterinary medicine to immobilize rodents and lagomorphs for various purposes, including:

Examinations and procedures: This allows for safe and thorough physical examinations, collection of diagnostic samples, and performance of minor surgical procedures.

Minimizing stress and injury: Restraint helps minimize stress and potential injuries to both the animal and the handler during handling procedures.

Treatment: Chemical restraint can be used to administer medications or treatments that would be difficult or dangerous to administer to a conscious animal.

However, it is important to note that chemical restraint should only be performed by qualified veterinarian or trained personnel under veterinary supervision. Improper use of these medications can have serious consequences for the animal, including:

Overdose: This can lead to respiratory depression, cardiac arrest, and even death.

Hypothermia: Chemical restraints can lower body temperature, so it›s crucial to maintain proper body temperature during and after restraint.

Allergic reactions: While uncommon, some animals may experience allergic reactions to certain medications.

5.5.1. Choosing the Right Restraint

The choice of chemical restraint agent depends on several factors, including:

Species: Different species have varying sensitivities to different medications.

Age and health status: Younger, healthier animals can generally tolerate a wider range of medications than older or sick animals.

Purpose of restraint: The type of procedure being performed will influence the level of restraint required.

5.5.2. Types of Chemical Restraint Agents

There are various classes of chemical restraint agents used for rodents and lagomorphs, each with its own unique properties and effects. Here are some commonly used agents:

Tranquilizers: These medications cause sedation and relaxation, making the animal calmer and easier to handle. Examples include acepromazine (ACP) and medetomidine.

Injectable anaesthetics: These medications induce a state of unconsciousness, allowing for the performance of more complex procedures. Example include ketamine.

Inhalational anaesthetics: These are volatile gases that are inhaled by the animal and induce unconsciousness. Isoflurane is a commonly used inhalational anaesthetic for rodents and lagomorphs.

Anaesthesia for rodents is a crucial procedure used to ensure their safety and well-being during various medical procedures, including:

Examinations and diagnostics: This allows for safe and thorough physical examinations, collection of diagnostic samples (blood, tissue), and imaging procedures (X-rays, ultrasounds).

Surgery: Anaesthesia is necessary for performing surgical procedures, even minor ones, to minimize pain and stress for the animal.

Treatments: anaesthesia can be used to administer medications or treatments that would be difficult or dangerous to administer to a conscious animal.

5.6. Anaesthesia for Rodents

There are two main types of anaesthesia used for rodents:

1. ***Inhalational anaesthesia:*** This method involves administering the anaesthetic agent through inhalation, typically via a nose cone or mask. Isoflurane and sevoflurane are the most commonly used inhalational anaesthetics for rodents.
2. ***Injectable anaesthesia:*** This method involves injecting the anaesthetic agent intramuscularly (IM) or intraperitoneally (IP). A combination of Ketamine and Xylazine is a common injectable aesthetic used for rodents.

5.6.1. Choosing the Right Anaesthesia

The selection of the most appropriate anaesthetic for a rodent depends on various factors, including:

Species: Different rodent species may have varying sensitivities to different anaesthetic agents.

Age and health status: Younger, healthier animals generally tolerate anaesthesia better than older or sick animals.

Length and complexity of the procedure: Shorter and less invasive procedures may require only mild anaesthesia, while longer or more complex surgeries may necessitate deeper anaesthesia.

5.6.2. Additional Considerations

Monitoring: During anaesthesia, the rodent's vital signs, such as heart rate, respiration, and body temperature, must be closely monitored to ensure their safety and well-being.

Pain management: Even under anaesthesia, rodents may still experience some pain. Therefore, it's essential to use appropriate pain medication (analgesics) before, during, and after the procedure.

Recovery: After anaesthesia, the rodent requires a warm and quiet environment for recovery until fully awake and mobile.

5.7. Anaesthesia for lagomorphs

Anaesthesia for lagomorphs, primarily rabbits, is similar to that of rodents but requires some specific considerations due to their unique physiological characteristics.

5.7.1. Types of anaesthesia

Inhalational anaesthesia: Similar to rodents, isoflurane and sevoflurane are the most commonly used inhalational anaesthetics for lagomorphs.

Injectable anaesthesia: Combinations like Ketamine-Xylazine or Medetomidine-Midazolam-Butorphanol are used for procedures requiring deeper anaesthesia or pain management.

5.7.2. Considerations for Lagomorphs

Stress management: Lagomorphs are highly susceptible to stress, which can negatively impact anaesthesia and recovery. Preoperative stress reduction techniques like quiet handling and providing familiar hiding spots are crucial.

Dental anatomy: Their unique dentition, with sharp and continuously growing incisors, requires extra care during intubation (placing a breathing tube) to avoid accidental damage.

Body temperature regulation: Lagomorphs have limited ability to regulate their body temperature. Maintaining warmth with heating pads or blankets is essential during and after anaesthesia to prevent hypothermia.

Fast GI tract: Their rapid digestive system can lead to regurgitation under anaesthesia.

Pain management: Similar to rodents, providing appropriate pain medication before, during, and after procedures is crucial for their well-being.

Endotracheal Intubation: Though endotracheal intubation is difficult, preference to use Cuffed endotracheal tube may be given over to the uncuffed ones. Laryngospasm is a common occurrence with endotracheal intubation. Supraglottic airway device can be a good substitute to avoid the complications of endotracheal intubation.

5.7.3. Choosing the Right Anaesthesia

As with rodents, several factors influence the choice of anaesthetic for lagomorphs:

Species: While rabbits are the most common lagomorph pet, other species like hares may require different protocols.

Age and health status: Younger, healthy animals generally tolerate anaesthesia better than older or sick ones.

Length and complexity of the procedure: Similar to rodents, the procedure›s nature determines the required anaesthesia depth.

Local anesthesia combined with manual restraints may be used to facilitate minor surgical procedures. Ketamine HCl, intramuscular 4.4 mg/100 gm in the smaller species, up to 11 mg/kg in the larger species can be used for minor surgery. Ketamine can be also be combined with promazine, acepromazine or xylazine, to give a smooth and longer immobilization with better muscle relaxation and smoother recovery.

Chemical agents used for anesthesia in Rodents and lagomorphs (Wallach and Boever, 1983)

Agent	Small species (mg/kg)	Large species (mg/kg)	Effects
Ketamine	30-40	8-10	Good immobilization
Phencyclidine	1-2	0.5-1.0	Convulsions occasionally
C_{1744}	7-10	5-10	Excellent immobilization
Pentobarbital sodium	30-70	30-50	Good immobilization
Fentanyl and droperidol	0.5-0.7	0.1-0.5	Good immobilization

- Phencyclidine + Promazine, Dose is 0.5- 1.0 mg/kg
- Pentobarbital sodium -60 mg/ml solution. Diluted 10 times with distilled water 0.85 ml of the dilute solution (100 g/mg body weight, administered intraperitoneally
- Combination of Droperidol (20 mg/ml) and Tentanyl Citrate (0.4 mg/ ml). It was found to be safe and reliable anesthetic agent for use in the

rat (Thayer *et al.,* 1972). Intramuscular injection of 0.016 ml/100 gm of body weight was sufficient for restraint. Intramuscular injection of 0.05 ml/100 gm of body weight produced sufficient anesthesia.

- Combination of ketamine and xylazine, ketamine, droperidol and fentanyl and sodium pentobarbital in prehadrons ground squirrel is used (Olson, 1986).
- In rats xylazine-ketamine induced anesthesia which is antagonized by yohimbine.

5.7.4. Inhalation anesthesia

Intubation is necessary for inhalation anesthesia. Intubation of the trachea is a challenging task and can lead to laryngospasms. Intubation can be done using cuffed or uncuffed endotracheal tubes and the preference should be to use cuffed ones. Supraglottic airway devices can be a good replacer for endotracheal intubation, which can save time and protect from laryngospasms. Oxygen flow rate may be adjusted between 0.8- 1.5 L/min for induction and for maintenance it can be reduced to 400-800 ml/ min. Isoflurane is a good choice of anesthetic and can be adjusted to deliver 3%-5% isoflurane.

General considerations for anesthesia in rodents and lagomorphs

Physiological support

a. Maintenance of body temperature: an external heating source should be provided. Hypothermia is a major concern in rodents and lagomorphs and body temperature should be monitored at regular intervals. To maintain the normal body temperature, an external heating device should be used intraoperatively till recovery.
b. Fluids: if the procedure is expected to last more than 30 minutes, fluids, through vascular access, should be administered. Isotonic fluids may be choice of administration if specific losses are not known.
c. Vascular access: the lateral ear vein (marginal ear vein), cephalic vein and recurrent tarsal vein can be selected for vascular access and the lateral ear vein is the preferred one. Administration of sedative/tranquilizer and topical anesthetic application at the phlebotomy site can reduce stress level and pain.

5.7.5. Preanesthetic preparations

As in any other animal planned to undergo anesthesia, history is very important in rodents and rabbits as well. Classify them according to their physical status. Preoperative preparation doesn't include withholding food in rabbits and guinea pigs, as this may have adverse effects related to gastrointestinal disturbances. A rodent or rabbit may be considered risky for anesthesia when presented with nasal discharge, increased respiratory sound, or effort. Animals with signs of dehydration, anorexia and obesity can be risk factors during anesthesia (Wenger, 2012). For the calculation of drugs and fluids, the weight of the animal should be accurately taken. Food may be withdrawn approximately 1 hour prior to the anesthetic procedure to reduce the presence of food inside the oral cavity. Sedatives when used as a preanesthetic can have a smooth anesthetic induction. Midazolam is a good choice and can have approximately 1 hour of sedative action and has minimal cardiorespiratory depression. Glycopyrrolate is the preferred anticholinergic agent which will reduce bronchial and salivary secretions. It would be better to preoxygenate animals with respiratory or heart diseases to maintain oxygen saturation. Anesthesia in rabbits shall not be induced only with inhalant anesthetic due to struggling and breath-holding which may cause prolonged periods of apnea, bradycardia, hypercapnia, and hypoxemia. Always prefer a balanced anesthetic protocol for anesthesia induction in rabbits.

Table: Routinely used chemical restraining agents in rodents and lagomorphs (Flecknell and Thomas, 2015)

	Mice	**Rats**	**Guinea pigs**	**Rabbits**
Acepromazine	2-5 mg/kg IP, SC	2.5 mg/kg IM, IP	0.5-1 mg/kg IM	1 mg/kg IM
Atropine	0.04 mg/kg SC	0.05 mg/kg IP, SC	0,05 mg/kg IM, IP	
Glycopyrrolate				0.1 mg/kg IM, IV
Diazepam	5 mg/kg IP	2.5-5 mg/kg IP	2.5 mg/kg IP, IM	1-2 mg/kg IM
Midazolam	2-3 mg/kg IM#	2.5 mg/kg SC, IM#	0.5-2 mg/kg IM#	0.5-2 mg/kg IV, IM
Ketamine	100-200 mg/kg IM	50-100 mg/kg IM, IP	40-100 mg/kg IM, IP	25-50 mg/kg IM
Xylazine	5-10 mg/kg IP	1-5 mg/kg IP	5 mg/kg IP	2-5 mg/kg IM, SC

Table: Commonly combined drugs used as injectable anesthetics in rodents and rabbits

Drug combination	**Rabbit**	**Guinea pig**	**Mouse**	**Rat**
Ketamine + Medetomidine	15-20 mg/kg + 0.25-0.5mg/kg	40 mg/kg + 0.5 mg/kg	50-75 mg/kg + 0.25 – 1 mg/kg	75 mg/kg + 0.5 mg/kg
Ketamine + Midazolam	15-25 mg/kg + 1 – 3 mg/kg	5-15 mg/kg + 0.5 – 1 mg/kg		5 – 10 mg/kg + 0.25– 0.5mg/kg
Tiletamine/ Zolazepam	3 mg/kg			5-10 mg/kg IM

5.7.6. Induction and maintenance

The recommended anesthetic circuits for rodents and rabbits are Bain circuit or Ayre's T-piece which have low dead space and low resistance. Facemasks are usually used in rodents for the delivery of inhalant anesthetics. Though endotracheal intubation is challenging in rabbits, it may be performed. blind technique by listening to the respiratory sounds or visualizing the larynx with a laryngoscope or endoscope may be done for intubation. The rabbit shall be in a sternal position with the neck hyperextended to facilitate intubation. Spraying of local anesthetics (2% Lidocaine) on the larynx can reduce the incidence of laryngospasm due to endotracheal intubation. Laryngeal masks and supraglottic airway devices are newer equipment to replace the endotracheal tubes in rabbits which reduces the challenges of endotracheal intubation and associated laryngospasms.

Rodents: Inhalation anesthesia can be achieved using an anesthetic chamber, mask or endotracheal tube. The anesthetic-soaked cotton balls or gauze shall not come in direct contact with the animals. The chamber technique can be used for very short-period procedures. Flow through anesthesia chambers and facemasks are preferred and require an anesthetic machine, vaporizer, and oxygen. Due to the small size of rats, a nonrebreathing circuit is used for anesthesia. Corneal injuries are common following long-term anesthesia in rodents which can be prevented by using ophthalmic ointments. Endotracheal intubation in rats is difficult. An intravenous over-the-needle catheter (14g-20G) can be used for endotracheal intubation.

Isoflurane is an inhalant anesthetic of choice in rodents. It has rapid and reliable recovery. Sevoflurane also has a quicker recovery time compared to isoflurane. Rodents and rabbits have an increased rate of anesthesia associated mortality (Brodbelt *et al.*, 2008).

5.7.7. Anesthesia Monitoring

The basic vital parameters, which include, temperature, respiratory rate, the color of the mucous membrane and pulse should be monitored during anesthesia. Preferably the monitoring should be done at an interval not more than 5 minutes. Oxygen saturation can be measured using a pulse oximeter placed on a foot or base of the tail. Reflexes, mainly palpebral reflex and pain reflex, inability to maintain its normal stance, inability to move, and relaxation of muscles can be used to assess the stages of anesthesia. Invasive blood pressure can be monitored via the central auricular artery in rabbits or non-invasively using cuffs on a limb.

Hypothermia is of major concern in rodents undergoing anesthesia. This can be correlated with a greater body surface area to body mass ratio. Hypothermia will delay the recovery and may be a cause of extra stress on rodents. Warm water blanket drape, patient heating device, warm air blanket, administering warm fluids etc can be used to counteract hypothermia.

5.7.8. Recovery from anesthesia

Continue monitoring rodents in the surgical room/ anesthesia room until recovery. Monitor every 5-10 minutes until they start moving. Feed shall be offered once rodents have recovered completely. The recommendation is to monitor the animal for the first 3 hours during recovery. Hypothermia being of greater concern, temperature should be monitored until complete recovery. Gastrointestinal motility can be reduced due to anesthesia associated stresses, so monitor feeding and defecation. It can be treated by assisted feeding and GI motility enhancers. Pain medications are to include in the post-op period. Coprophagia is a normal activity and shall not prevent by using Elizabethan collars.

5.7.9. Local anesthesia

Regional and local anesthetic techniques can also be performed in rodents and rabbits. The commonly used local anesthetics are lidocaine and bupivacaine. These local anesthetics can be used for topical application, tissue infiltration, intraarticularly, regional nerve block and epidural analgesia. Lumbo-sacral space is the preferred site for epidural analgesia due to the relatively large space between L7 and S1 (Greenaway *et al.*, 2001). The dose of 0.1 – 0.2 ml/kg of local anesthetic agent can be used for epidural analgesia.

6

Chemical Restraint of Carnivores

Carnivora is an order of placental mammals that have specialized in primarily eating flesh, whose members are formally referred to as carnivorans. The order Carnivora is the fifth largest order of mammals, comprising at least 279 species. They come in a very large array of different body plans with a wide diversity of shapes and sizes. The common characteristics of carnivores are given below:

- They are meat eaters.
- There is the presence of canine teeth.
- The intestinal tract is short and is adapted to rapid digestion as well as the assimilation of meat.
- All species have anal glands. Some species, like the striped skunk (*Mephitis mephitis*), may eject the contents of the anal gland as a defensive maneuver.
- Os penis is present in males.
- There is a lack of a clavicle which helps in the freedom of movement of the forelimb.
- Ulna is well developed.
- Toes end in claws.

Carnivora can be divided into two suborders: the Feliformia, containing the true felids and several «cat-like» animals; and the Caniformia, containing the true canids and many somewhat «dog-like» animals. The feliforms include families such as the felids or cats (both great cats and lesser cats), hyenas, mongooses, and civets. The caniforms include the canines, bears, raccoons, mustelids, skunks, and pinnipeds. Members of this group are found worldwide with immense diversity in their diet, behavior and morphology.

6.1. Classification of Carnivores

Family	Common names	No. of species
Canidae	Wild dogs, jackal, fox, wolf	37
Felidae	Tiger, lion, panther, jungle cat etc.	36
Ursidae	Bears	7
Mustelidae	Skinks, otters, weasels	68
Viverridae	Mongooses, Civets	82
Procyonidae	Raccoons, kinkajou, pandas	18
Hyaenidae	Hyaenas	4

The genus Felis is a group of small and medium-sized cat species native to Africa, Europe, and Asia. The most famous member of this genus is the domestic cat (*Felis catus*), but there are also six wildcat species:

- The jungle cat (*Felis chaus*) is the largest member of the Felis genus, with a head-and-body length of 62-76 cm (24-30 in) and a tail length of 25-35 cm (9.8-13.8 in). It is found in grasslands, swamps, and forests from India to Southeast Asia.
- The black-footed cat (*Felis nigripes*) is the smallest wildcat in Africa, with a head-and-body length of 38-42 cm (15-17 in) and a tail length of 15–20 cm (5.9–7.9 in). It is found in the dry savannas and grasslands of southern Africa.
- The sand cat (*Felis margarita*) is a small cat found in the deserts of North Africa and the Middle East. It has a head-and-body length of 45-57 cm (18-22 in) and a tail length of 25-35 cm (9.8-13.8 in).
- The African wildcat (*Felis lybica*) is the ancestor of the domestic cat. It is found in Africa, Europe, and Asia. It has a head-and-body length of 47-78 cm (19-31 in) and a tail length of 27-39 cm (11-15 in).
- The European wildcat (*Felis silvestris*) is found in Europe and western Asia. It has a head-and-body length of 49-78 cm (19-31 in) and a tail length of 25-35 cm (9.8-13.8 in).
- The Chinese Mountain cat (*Felis bieti*) is found in the mountains of central China. It is the least-known member of the Felis genus. It has a head-and-body length of 60–80 cm (24-31 in) and a tail length of 30-40 cm (12-16 in).

The others are Mountain lion (*Felis concolor*), Jaguarundi (*Felis yagouaroundi*), Steppe Cat (*Felis manul*), Serval (*Felis serval*), Leopard Cat (*Felis bengaleneses*), Rusty-spatted Cat (*Felis rubiginosa*), Fishing Cat (*Felis viverrine*), Flat headed Cat (*Felis planiceps*), Marbled cat (*Felis marmarata*), Golden Cat (*Felis aurata*) Ocicat (*Felis paradalis*), Margay (*Felis wiedil*),

Little Spotted cat (*Felis tigrine*), Pampas Cat (*Felis pajeros*), and Geoffray Cat (*Felis geoffroyi*).

6.2. Felids

The Genus Panthera includes:

- Asiatic lions (Panthera leo) are found only in the Sasan Gir National Park of Gujarat's Saurashtra Peninsula, covering about 1412 square kilometers. These are the majestic animals in Indian forests, living in pride.
- Tigers (Panthera tigris) are endangered wild animals found all over India, from the Himalayas to Cape Camorin, except in the deserts of Rajasthan, Punjab, Kachchh, and Sindh.
- Panthers (Panthera pardus) are highly adaptable in nature to the environment and are seen all over India. There are three races present in India (Panthers from Sindh, Baluchistan, and Kashmir are regarded as separate races).
- Jaguars (Panthera onca) are a large cat species and the only living member of the genus Panthera native to the Americas. With a body length of up to 1.85 m (6 ft 1 in) and a weight of up to 158 kg (348 lb), it is the biggest cat species in the Americas and the third largest in the world.

6.3. Other Felids

- Clouded Leopard (*Neofelis nebulosa*): seen in Assam and Sikkim.
- Asiatic cheetah or Hunting Leopard (*Acinonyx jubatus*): extinct in India at present.
- Caracals and Jaguars: Caracals are seen in the north and north-west hills of Kachchh, and Jaguars are considered to be sturdy animals.
- Lynx (*Felis lynx*): seen in the upper Indus valley, Ladakh, Gilgit, and Tibet. This animal is called "isabellina".
- Indian desert cat: seen in deserts of north-western India extending to the drier regions of Central India.
- Marbled Cat (*Felis marmorata*): sighted in Sikkim and Assam. A single race occurs in India.
- Leopard Cat (*Felis bengalensis*): sighted in wider parts of India, from Kashmir to Cape Camorin.
- Golden cat (*Felis temmincki*): seen in Assam and Sikkim.
- Pallas Cat (*Felis manul*): seen in Ladakh.

6.4. Immobilization Techniques of Felids

Precautions: In contrast to our domestic animals, precautions such as fasting and preventing drinking are not possible for free-ranging and many of our captive felids. When possible, 24-48 hours of fasting should be performed. Preventing access to water for 4-6 hours prior to immobilization is also desirable. Whenever possible, immobilization should take place in a relatively small, confined area, away from water hazards and areas where animals could fall, i.e., moats. Hand injection of immobilization drugs is much less stressful to the animal than darting and can frequently be performed in small cages or squeeze cages and/or with training. For years, ketamine HCl and an alpha-2 adrenergic antagonist (aka, alpha-2) have been the principal drugs used for immobilization or chemical restraint of large felids (Table 1).

Table 1. Immobilization and emergency drug dosages for lions, tigers and leopards.

Species	Agents	Reference
Lions, Tigers and Leopards	Medetomidine (20-30 μg/kg) or Dexmedetomidine (10-15 μg/kg) and Midazolam (0.1 mg/kg) & Ketamine (2.0-3.0 mg/kg); Intramuscular administration	Edward C. Ramsay (2020) Author's personal experience
Amur tiger (snared)	Ketamine (10.8 mg/kg) and Xylazine (0.81 mg/kg); Intramuscular administration	Goodrich et al (2001)
Amur tiger and Amur leopards	Ketamine (6.6 mg/kg) and Xylazine (0.66 mg/kg); Intramuscular administration	Quigley et al (2001)
Leopards (Wild)	Ketamine (2.2-2.6 mg/kg) and Xylazine (1.1-1.3 mg/kg); Intramuscular administration	Sontakke et al (2009)
Reversal Drugs		
Yohimbine	1.25 mg/kg; Intramuscular or Subcutaneous administration	
Atipamezole	5 mg/mg Medetomidine or 10 mg/mg Dexmedetomidine; Intramuscular administration	
Emergency Drugs		
Doxapram	3-5mg/kg; Intramuscular or Intravenous administration	
Midazolam	0.1 mg/kg; Intravenous administration	
Analgesic Drugs		

Meloxicam	0.1-0.2 mg/kg Subcutanously (initially), then 0.1 mg/kg, Post-operatively after 24 hours	
Tramadol	1.0-2.0 mg/kg Post-operatively, twice a day	
Buprenorphine	0.03 mg/kg; Subcutaneous administration	

While ketamine has no reversal agent, it is a relatively safe and short-acting drug. Ketamine, as sold in most countries, is unconcentrated (either 50 mg/mL or 100 mg/mL), and, as such, large volumes may be required for large felids. Concentrated ketamine (200 mg/mL) can reduce the size or number of ketamine injections or darts required, but the concentrated formulation is not easily available. Ketamine is fairly quick-acting, with signs of ataxia occurring within 5 minutes of intramuscular injection, but, depending on dosage, effects may begin to 'wear off' 30-40 minutes following administration. Redosing with ketamine, either intramuscular or intravenous, is frequently required for performing long procedures (Edward C. Ramsay, 2020).

Ketamine at high dosages has the potential to cause seizures, which occur most frequently in tigers, so a benzodiazepine, such as diazepam or midazolam, should always be available or added to the initial drug combinations. If 48 tigers had seizures during a past immobilization, they have an increased chance of doing so in future immobilizations. Diazepam has variable intramuscular absorption and should be given either orally or intravenously. Oral administration of diazepam, pre-darting, however, has not been successful in preventing seizures. Rather, for animals prone to seizures, reduce the dosage of ketamine to just what is required by using a larger dosage of the alpha-2 agent and adding midazolam, which is absorbed well intramuscularly, to the initial immobilization injection (Edward C. Ramsay, 2020).

Telazol,™ or Zoletil,™ is a combination of tiletamine, a dissociative agent, and zolazepam, a benzodiazepine, and has been used by many veterinarians for felid immobilizations. When reconstituted according to directions, it is relatively concentrated (50mg of each drug/mL), allowing for small injection volumes. It also has the advantage of being able to be reconstituted with less water or another immobilization agent, i.e., xylazine, making the initial injection volume even smaller. Early work suggested Telazol™ should be avoided in tigers, but a review of those papers refuted that recommendation, and it is now considered safe to use in tigers (Kreeger and Armstrong, 2010). The major complication I have seen with Telazol™ is occasional, very long recoveries; some animals take over 6 hours to recover. These long recoveries seem to occur randomly but appear more frequent when Telazol™ is used in combination with other drugs or is redosed. Due to this later concern, if an animal has received Telazol™ as the initial immobilization agent, should the animal need more sedation, don't redose it with more Telazol,™ but use ketamine instead. Similar to ketamine, tiletamine is not reversible. Zolazepam can be partially reversed with flumazenil, but the dosage of flumazenil is large, flumazenil is short-acting, and it is relatively expensive. As such, I only use flumazenil in animals experiencing exceptionally long recoveries (Edward C. Ramsay, 2020).

Xylazine has been the most utilized and least expensive alpha-2 immobilization agent for large felids. It comes in a concentrated, "large animal" preparation (100 mg/mL), it is inexpensive, and its effects can be reversed with yohimbine. Yohimbine is marketed for small animals in a dilute concentration, but concentrated forms may be obtained from compounding pharmacies. It should be given intramuscularly unless there is an emergency, and even then, full reversal doses should not be given intravascularly due to their profound cardio-vascular effects. Xylazine should not be reversed until at least 40 minutes after initial ketamine injection to avoid ketamine 'excitement' following the alpha-2 reversal (Edward C. Ramsay, 2020).

Xylazine has profound cardio-vascular effects and has been replaced in our practice by medetomidine or dexmedetomidine. The latter is the bio-active enantiomer of medetomidine. As they are marketed for small animals, these drugs are not concentrated, but concentrated forms can be obtained from compounding pharmacies. The effects of both medetomidine and dexmedetomidine can be reversed with atipamezole. All three of these drugs are expensive. There are two adverse effects seen with medetomidine and dexmedetomidine: vomiting and shallow breathing. Vomiting during induction is more common and typically occurrs within 5-10 minutes after injection in 49 animals with these agents (up to 30% of the animals). The animal may still be standing at this time and aspiration of vomitus has not occurred, in my experience. More concerning is bradypnea: shallow breathing with deep breaths occurring only once or twice per minute. Counterintuitively, this appears to occur more often in tigers receiving low dosages of medetomidine. When low respiratory rates or mucous membrane color cause concern, respiration can be stimulated with doxapram intramuscular or intravenous or, if in a safe environment, a small amount (approximately 10-15% of the total reversal dose) of atipamezole can be given intravenous. I usually administer the latter into the lingual vein (Edward C. Ramsay, 2020).

Large felids seem to breathe best in lateral recumbency. During short immobilization procedures, endotracheal intubation is not typically required, in part because if the animal is not sufficiently sedated with an alpha-2 drug, it may 'fight the tube,' if one attempts to intubate it. Monitoring during immobilization usually involves recording respiratory rates, mucous membrane color, heart rate, and rectal temperature. If a pulse oximeter is available, the monitoring of SpO2 is also valuable. Supplemental oxygen should be available for animals with reduced respiratory rates and SpO2<90%. Oxygen can be supplied by nasal insufflation at 2 liters per minute. I keep doxapram (for apnea), midazolam (for muscle relaxation and for prevention or treatment of seizures), and reversal agents available as 'emergency drugs' for all immobilizations (Table 1) (Edward C. Ramsay, 2020).

6.5. Anesthesia

For longer procedures, such as dental procedures or surgeries, full anesthesia is required. Felids are typically induced using one of the immobilization regimens described above. If the depth of immobilization is sufficient, then endotracheal intubation can be performed with injectable drugs alone. If the animal is not sufficiently sedated for intubation, then it can be further relaxed by administering an inhalant agent via face mask. Most large felids do not require administration of topical lidocaine to assist or avoid laryngospasm, but it is necessary to dramatically extend the animal's neck- to almost a 90° angle to the thoracic spine- to accomplish intubation. The use of a rigid stylet within the endotracheal tube is also helpful. Most tigers will take an 18 mm or 20 mm endotracheal tube, but larger felids may need a 22 mm tube.

Anesthesia is maintained in our practice with isoflurane in oxygen, typically ranging from 2% to 3%. Most animals breathe well under anesthesia, but partial reversal of the alpha-2, typically 10-20% of the total atipamezole reversal dose intramuscular, should be administered to improve respiration if needed. Monitoring during anesthesia is similar to that described above, but may also include monitoring end-tidal CO2 and indirect blood pressure monitoring. A wide range of end-tidal CO2 readings can be tolerated, but should CO2 levels exceed 55 mmHg, manual or mechanical ventilation should be instituted. Similarly, mean blood pressures (MAP) can vary widely during anesthesia, but long periods of low MAP (<50 mmHg) should be addressed via increased administration of fluids, reducing the inhalation agents (possibly administering intravenous ketamine to maintain safe restraint), and/or the administration of vasopressors.

A major complication of long anesthesia (>2 hours) is the development of hyperkalemia (K+ > 5.5 meq/L). Most animals given alpha-2 agents will show gradually increasing blood glucose concentrations and concomitant increases in plasma K+ levels during anesthesia. Unfortunately, for reasons yet to be discovered, this occurs in some individuals at a very rapid rate, resulting in life-threatening hyperkalemia prior to the end of the surgery or procedure. Early administration of atipamezole reduces the rate of K+ increases, and as a result, we routinely administer 12.5-16% of the full atipamezole reversal dose intramuscularly, 2 hours after the initiation of inhalation anesthesia. Repeat that atipamezole dose every 30 minutes through the remainder of the surgery or until the full, calculated reversal dose has been given (Edward C. Ramsay, 2020).

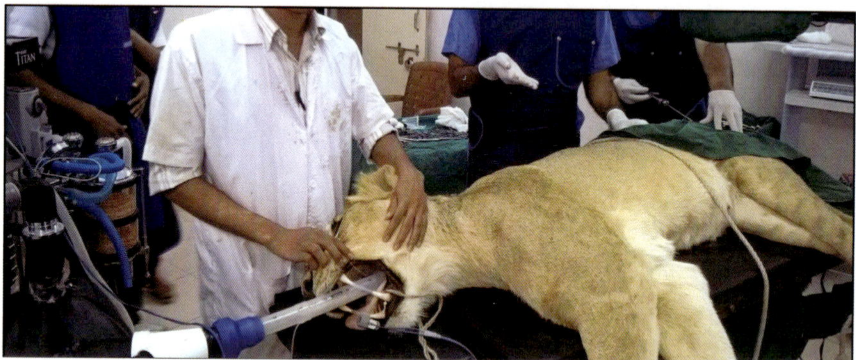

General anesthesia in a Lion

6.6. Analgesia

For acute pain, such as traumatic injury or post-surgical pain, we rely on a combination of oral non-steroidal inflammatory drugs (NSAIDs) and opioids. Meloxicam is most commonly used NSAID, and the initial dose is given during surgery, followed by once-daily dosing for 3-7 days (Table 1). Pre- and post-operative tramadol is given for minor procedures (i.e., castrations and endodontics), and continued twice a day postoperatively, as needed. For more painful procedures, such as pyometra surgeries, buprenorphine (subcutaneously, intra-operatively) is used instead of tramadol on the day of surgery. A single dose is usually sufficient, but an additional dose can be administered 8-12 hours after the initial dose, if necessary. Typically, we transition to oral tramadol the day following surgery. A number of opioid analgesic applications, such as epidural morphine, sustained-release fentanyl patches, and long-acting buprenorphine preparations, have gained popularity in small animal practice. These are frequently expensive and have not seemed to be of great benefit. Additionally, the use of multiple, simultaneous opioids may result in opioid dysphoria. A tiger experiencing this has become manic and is in danger of injuring itself. The condition is treated, and the diagnosis is confirmed, by administration of naloxone or naltrexone intramuscularly, with immediate effect (Edward C. Ramsay, 2020).

Oral tranquilizers offer little restraint for examination or treatment and are sometimes helpful when introducing animals to each other. The dose is 2-6 mg/kg of body weight. Phencyclidine and ketamine also have some efficacy. But the effects of these drugs give variable results when administered orally.

Intramuscular chemical agents are most commonly used on wild felines. Tiletamine HCl has been reported (Krahwinkel, 1970) as a useful anesthetic agent in lions, combined with atropine sulfate at 0.02 mg/kg and tiletamine HCl at 1-1.25 mg/kg of body weight. Anesthesia was maintained by methoxyflurane and a combination of 50% N_2O and 50% O_2 in a semi-closed system. This dosage used in lions is much less than that used in domestic cats, but slightly higher than that used in tigers and leopards. To get sufficient analgesia and muscular relaxation for major surgical procedures, it is necessary to supplement this agent with other anesthetics.

For more than a decade, phencyclidine HCl was considered the anesthetic of choice for the immobilization of lions and leopards. But it had many disadvantages. It caused muscular spasms and convulsions, which, at high ambient temperatures, contributed to hyperthermia. There is a loss of thermoregulation, the duration of anesthesia is long and there are no antagonists. The recovery period was protracted, and animals were guarded for many hours.

Tiletamine and Zolazepam were used for the immobilization of wild lions and leopards. There is a linear relationship between the duration of anesthesia and the dose. Lions more susceptible than leopards, and males are more susceptible than females. There is a 15-minute longer recovery in males compared to females (King *et al.,* 1978).

Ketamine is a unique feline anesthetic agent. The dosage varied from 5.3-56.7 mg/kg body weight (Beck and Ott, 1971). Light-weight darts and a blow gun were found to be useful as a supplement to longer-range dart projector systems. Ketamine alone causes muscular rigidity and even convulsions. A combination of xylazine HCl and ketamine HCl at rates of 55 mg/mL and 200 mg/mL, respectively, was used in African lions (Hrbst *et al.,* 1985). A dose rate of 110 mg xlyazine and 450 mg ketamine HCl was used. These doses could be delivered easily in 3 ml darts. The side effects of ketamine alone were overcome by this combination.

Phencyclidine HCl was used in the African lion intramuscularly at a rate of 0.5-2.2 mg/kg of body weight, and the duration was 30-60 minutes. Associated termers can be eliminated by administration of promazine HCl (Sparine) at a rate of 0.5-1.0 mg/kg body weight. In place of promazine, acepromazine can also be used (Campbell and Harthoorn, 1963).

Tilazol is a 1:1 combination of Tiletamine HCl (CI-634) and Zolazepam HCl (CI-716) to overcome the adverse effect of Tiletamine alone. Inhalation anesthesia can be administered by using halothane, N_2O, or methoxyfurane in combination, either alone or after induction by other agents, maintained by inhalants.

Ketamine and xylazine combination was used to immobilized Afroasciatic lion. A darting pistol (Dist. Inject Pistol, Model -30) was used with blue cartridge. A 5 ml metallic syringe (dart) with a barbed needle was filled with 3 ml Ketamine HCl (100mg/ml) and 2 ml Xylazine (100mg/ml). The animal was immobilized within 15 minutes and was shifted. Yohimbine HCl (Antagozil Injection-Troy Laboratory Private Limited, Australia) 3 ml was administered intravenously as antidote and animal recover within 3 minutes (Bonal *et al.*, 1994).

Sabapara (1989) used 1000mg of pure Ketamine to translocate the lions. For darting Dist. Inject M-60 mode gun was used. In 5 animals 1000mg of Ketamine was used and in another two animals additional dose of Ketamine plus Rompun was used.The imbalance time ranges from 8-32 minutes, animals were completely unconscious in 8-45 minutes and recovery time ranges from 10-50 minutes.

K. Chandrasekara Pillai (1992) used 200mg of Ketamine and 40 mg of Xylazine in Asiatic lion for the examination and treatment of swelling paws. The animal was completely immobilized after 18 minutes of injection. Additional 200mg of Ketamine HCl was given. Complete recovery occurs at 107 minutes and animal was able to walk.

6.7. Canidae

Canidae (from Latin *canis*, "dog") is a biological family of dog like carnivorans, colloquially referred to as dogs, and constitutes a clade Canidae is a family of mammals in the order Carnivora, which includes domestic dogs, wolves, coyotes, foxes, jackals, dingoes, and many other extant and extinct dog-like mammals. A member of this family is called a canid; all extant species are a part of a single subfamily, Caninae, and are called canines. They are found on all continents except Antarctica, having arrived independently or accompanied human beings over extended periods of time. The family includes three subfamilies: the Caninae, the extinct Borophaginae, & Hesperocyoninae. The Canidae are known as canines and include domestic dogs, wolves, coyotes, foxes, jackals, and other species.

A member of this family is also called a canid. Canids are the most commonly sighted wild animals in general. The offspring are born relatively underdeveloped and are very dependent on their parents for their survival. Many times, some species of canids travel a long distance as is the case with the wild dogs or jackals, in search of food resources.

6.8. Common Features of Canids

- Erect ears.
- Strong and non-retractile claws.
- Long muzzle.
- Males have a baculum in the penis.
- Perfect digitigrade feet.
- Multiparous in nature.
- All the canids almost resemble the domestic canids in terms of anatomical features.

Indian Wild Dog

- Indian wild dogs (*Cuon alpinus*) are also called dholes. These animals have a hunting habit in packs only. They prey on a number of animals, like deer, sheep, gaur, pigs, etc.
- A large pack may attach bigger animals like buffalo and gaur. They help to improve the prey population by eliminating old and diseased individuals.
- These are found in Andhra Pradesh and in the Mudumalai and Anaimalai regions.

Exotic Wild Dogs

- Exotic wild dogs are given below:
- Dingo (*Canis dingo*), or wild dogs of Australia.
- Cape hunting dogs of the south of the Sahara.

Jackal

- Golden Jackals (*Canis aureus*) are present throughout India in small numbers in any kind of habitat, ranging from humid, dense forests to dry, open plains. This animal comes out at dusk and retires by dawn. The other is the black backed jackal (*Canis mesometas*).
- This can make a typical howl that is long-drawn and high-pitched. Dead carcasses are eaten by them, and jackals also feed on weak livestock and poultry.

Indian Fox

- Indian foxes (*Vulpes bengalensis*) are the small, slim animals with slender limbs.
- This are seen in agricultural fields and are solitary hunters, but they appear to tolerate the presence of common mongoose near their den.

Red Foxes

- These animals are seen in Sikkim to the western Himalayas, including the arid zone of the north-west.
- In the cooler Himalayan region, the animals prefer to live amidst small, cultivated lands, and the animals are mainly nocturnal in nature.
- These animals' pair for life, occupying the same den year after year.

Wolf

- Wolf (*Canis lupus*) is seen in several parts of India.
- These animals assume a height of about 65-75 cm, and the weight of these carnivores may be about 18- 28 kg.

Hyaena

- Hyaena belongs to the family "Hyaenidae". These are the animals with powerful jaws. Striped hyaena (*Hyaena hyaena*) is a species present in India. Hyaenas have large anal glands and are scavengers.
- They are nocturnal in nature and are stocky dog like animals inhabiting the plains of southwest Asia and Africa. Walks on toes, four on each foot.
- There are three species of hyaena: striped hyaena (*Hyaena hyaena*), brown hyaena (*Hyaena brunea*), and spotted hyaena (*Crocuta crocuta*). (Note: Striped hyaena is the only species available in India).

6.9. Immobilization of canines

The chemical immobilization of the exotic canid is almost the same as that of the domestic dog, with minor exceptions. Oral tranquilizers commonly used on dogs, such as acepromazine, can be used. Tranquilizers are helpful for introducing animals to each other, but tranquilizers alone offer little restraint for examination or treatment. Intravenous barbiturates can be used if the animal is tractable; this is usually not practiced with wild canids. Ketamine hydrochloride at 10 mg/kg and xylazine HCl at 2 mg/kg intramuscular injections are given either by hand syringe once the animal is on the snare pole or by projectile syringe. Induction occurs within 3-5 minutes following the intramuscular injection. The duration of analgesia and anesthesia lasts for 15-20 minutes, and a complete recovery requires 60-90 minutes.

Ketamine can be used alone at higher doses (20-35 mg/kg of body weight); however, salivation, muscular rigidity, and even convulsions sometimes occur. The anesthesia induced by ketamine hydrochloride in conjunction with propiomazine resulted in a slightly increased duration and depth of anesthesia over that reported for ketamine and xylazine. The mean time of onset of anesthesia was 53 seconds after injection, with a mean dosage of 13.2 mg/kg of body weight. The recovery time is 80-115 minutes (Hallett *et al.*, 1979). Xylazine HCl, at 1-2 mg/kg of body weight, produces a level of sedation ranging from mild depression to immobilization.

Xylazine HCl at doses of 2.7-3.9 mg/kg of body weight was administered intramuscularly to wolves. The induction time is 2.5 minutes, and the duration lasts for 30-60 minutes. The recovery time is 1.7 hours (Michaelphilo 1978). Dogs can be safely and completely immobilized with a combination of xylazine HCl at 2.2 mg/kg and atropine at 0.5 mg/kg of body weight intramuscularly, respectively. The induction time is 3-13 minutes. It is antagonized by 4-aminopyridine (4AP) (0.3 mg/kg IM) and yohimbine (0.125 mg/kg IM) or a combination of both after injection (Walner *et al.*, 1982).

Kotwal et al (1991) used 0.75 ml of Hellabrum mixture (HBM) (125 mg xylazine and 100 mg ketamine/ml) for immobilization and radio-collaring of Golden Jackal (Canis aureus Linn). The animal became immobilize after 3 minutes of injection. One ml of Effortil was injected 30 minutes after darting and animal became mobile and gradually moved away after 5 minutes.

Drugs	**Mean times to ability walk**	**Mean time for total recovery**
Xylazine HCl without antagonists	75 min	3.8 h
4- aminopyridine antagonists	25.4 min	2.5 h
Yohimbine antagonists	8.7 min	1.1 h
Yohimbine + 4 AP antagonists	4.8 min	1.6 h

A combination of fentanyl and droperidol (Innovar-vet) produces good immobilization and analgesia in exotic canids at a dose of 1 ml/35 kg intramuscularly or 1 ml/75 kg intravenously. The effect lasts for 30-40 minutes. To minimize side effects like bradycardia and salivation, atropine should be given. Gas anesthesia, the same as that of domestic canids, can also be administered (Wallach and Boever, 1983).

Immobilizing agents for Canidae

Agents	Method of administration	Dose	Induction (minutes)	Duration (minutes)	Recovery (minutes)	Comments
Acetyl promazine	Oral	0.5-2.0 mg/ kg	30-60	60-120	?	Tranquilization only, no immobilization
Acetyl promazine	Intramuscular	0.5-1 mg/kg	10-15	60-90	?	Tranquilization only, no immobilization
Acetyl promazine	Intravenous	0.5-1 mg/kg	3	60-90	?	Tranquilization only no immobilization
Inovar vet	Intramuscular	1ml/ 35 kg	10-15	30-40	90-120	Good analgesia, has advantage of quick reversible
Inovar vet	Intravenous	1 ml/ 75 kg	3	30-40	90-120	-do-
Ketamine HCl	Intramuscular	20-35 mg/kg	8-12	15-20	60	Good analgesia and immobilization, convulsion seizers when used alone
Librium	Oral	?	30-60	60-120		Tranquilization only
Pentobarbital	Intravenous	?	1-2	15-20	60-90	Good analgesia and good muscle relaxation but has to be given intravenously
Xylazine HCl	Intramuscular	1-2 mg/kg	10-15	30-90	90	Tranquilization only will more when touched
Xylazine HCl	Intravenous	1-2 mg/kg	3	30-90	90	Tranquilization only will more when touched
Ketamine HCl + xylazine HCl	Intramuscular	10 mg/kg + 2 mg/kg	5-10	15-20	60-90	Best immobilization, good analgesia, good muscle relaxation, good safety margin, best overall general anesthetic

6.10. Anesthesia in Viverids and Hyaena

Viverrids belong to the family Viveridae. Viverrids have four or five toes on each foot and half-retractile claws. They have six incisors in each jaw and molars with two tubercular grinders behind in the upper jaw, and one in the lower jaw. The tongue is rough with sharp prickles. A pouch or gland occurs beneath the anus, but there is no cecum. Examples are African civets (*Civettictis civetta*), Indian civets (*Viverra zibetha*), Binturong or Bear cats (*Arctictis binturong*), Genets (*Genetta species*), Mongooses (*Herpestes species*), and Fossa (*Cryptoprocta ferox*).

Chemical restraint for Viveridae and Hyenas is very similar to that used for other members of the carnivore family. Ketamine HCl and xylazine HCl, at 10 mg/kg and 2 mg/kg body weight, respectively, are the drugs of choice for immobilization. Dosage levels of up to twice that amount can be used for a surgical plane of anesthesia. Ketamine alone, phencyclidine, and etorphine have been used, but they are less desirable. Intravenous barbiturates at similar dosages for domestic dogs and cats can be used. Inhalant anesthetics like methoxyflurane and halothane may be used to maintain anesthesia. Small viverids can be put in a bell jar with the halothane (Wallach and Boever, 1983).

Hyeanas

Etorphine-Methotrimeprazine, Dosage: 0.1 ml/kg, intramuscular administration

Ketamine-Xylazine, Dosage: 10.0 mg/kg; 2.0 mg/kg, intramuscular administration

Viverids

Phencyclidins, Dosage: 1 mg/kg, intramuscular administration.

Tiletamine + Zolazepam (C1744) Dosage: 4 mg/kg, intramuscular administration (Green, 1982).

Hellabrum mixture (HBM) (125 mg xylazine and 100 mg ketamine/ml) was used to immobilized Hyena (Hyaena hyaena). Initially 1ml HBM was injected using dart and it was repeated after 40 minutes. After 6 minutes of second injection the animal was completely tranquillized for treatment of severe infected wound. The animal was completely recovered with in 2 hours and started moving (Gopal, R., 1991).

George and Rajankutty (1990) used 3 ml of droperidol-fentanyl mixture intramuscularly (Innovar Vet Pitman Moore, Inc., Washington Crossing USA, 1ml contains 20 mg of droperidol and 0.4 mg of fentanyl citrate). The second dose of 1.5 ml was administered after 21 minutes of first injection. The animal assumed the sternal recumbency 3 minutes after second injection, followed by lateral recumbency and fully anaesthesia.

6.11. Immobilization of Procyonids

Procyonidae is a New World family of the order Carnivora. It includes raccoons (*Procyon lotor*), ringtailed cats (*Bassariscus astutus*), cacomistles, coatis, kinkajous, olingos (*Bassaricyon gabbii*), and olinguitos. Procyonids inhabit a wide range of environments and are generally omnivorous. Though the 18 species are classified as carnivores, procyonids are actually omnivorous and closely related to bears (family Ursidae). Indeed, both the giant panda (*Ailuropoda melanoleuca*) and the lesser, or red, panda (*Ailurus fulgens*) have been grouped in the past as procyonids; however, the giant panda is actually a bear, and the lesser panda is the sole member of the family Ailuridae. Procyonids mostly inhabit Central America, and only the North American raccoon is widely distributed north of the tropics. The raccoon and some other procyonids are sometimes kept as pets.

Chemical immobilization

Chemical restraint and anesthesia techniques and protocols applicable to captive mammals in the families Mustelidae and Procyonidae are similar in many respects to the techniques used for domestic dogs and cats. Numerous general and specific single drugs or drug combinations have been reported for chemical restraint, immobilization, and anesthesia of procyonids and mustelids.

The best chemical agents for immobilizing members of Procyonidae are a combination of ketamine HCl at 10 mg/kg and Xylazine @2 mg/kg of body weight, administered together intramuscularly. The intramuscular injection is given with a hand syringe once the animal is restrained in a net, on a snare pole, or in a squeeze cage. The induction occurs within 3-5 minutes. The duration of anesthesia is 15-20 minutes, and the complete recovery requires 60-90 minutes.

Ketamine HCl alone can be used at a rate of 20-30 mg/kg of body weight. The induction time is 7minutes, full recovery can be expected in 45-90 minutes. For deep anesthesia, the dosage can be increased up to 30-50 mg/kg of body weight.

Droperidol-Fentanyl combination (Inovar-vet) can be used intramuscularly at the rate of 1 ml/10 kg body weight.

Phencyclidine HCl and promazine HCl can be administered simultaneously by the intramuscular or intravenous route at a rate ranging from 0.5 to 2.0 mg/kg of body weight for clinical effects ranging from tranquilization to deep surgical anesthesia. The analgesic effect lasts from 60 to 180 minutes.

Tilazol, a 1:1 combination of Tiletamine HCl (CI-634) and Zolazepam HCl (CI-716), can be used for restraint and minor surgical procedures. The dose is 10 mg/kg of body weight given intramuscularly. The induction time is 3-10 minutes. The duration is 20 to 60 minutes, although surgical anesthesia is produced. Several reflexes persisted, including corneal, palpebral, pinnal, pedal, pharyngeal, and laryngeal.

Inhalation anesthesia is a lengthy procedure. An open non-rebreathing system is used most frequently because a small tidal volume of both halothane and methoxyflurane is used (Wallach and Boever, 1983). Raccons weighing 15 to 25 lbs are immobilized with capture equipment for approximately 7 minutes with 10 mg of succinylcholine chloride. Recovery is complete within 10 minutes (Lumb and Jones, 1984).

6.12. Immobilization of Mustelids

The family *Mustelidae* comprises stoats, polecats, mink, fishers, wolverines, weasels, martens, badgers, and otters and is the largest family within the order *Carnivora*. Mustelids have a global distribution, with members found on every continent except Australia and Antarctica. Mustelids are free-ranging, cover terrestrial and marine environments, are kept as pets, farmed, used as laboratory animals, and kept in zoos. Mustelids contain around 55–60 species across eight subfamilies. Skunks were once members of this family but have been recategorized into the family *Mephitidae* (Williams *et al.*, 2018). They included striped skunks (*Mephitis mephitis*), hooded skunks (*Mustela macroura*), spotted skunks (*Spilogale putoris*) and hog-nosed skunks (*Conepatus mesoleucus*).

Mustelids have existed in North America since the early Oligocene period (Paterson *et al.*, 2020). There are currently 11 species of mustelids that are native to North America, which include: American Badger (*Taxidea taxus*), American Marten, American Mink, Black-footed Ferret, Ermine, Fisher, Least Weasel, Long-tailed Weasel, River Otter, Sea Otter, and Wolverine. In Europe, mustelid species also include the stoat, the European polecat, the Eurasian badger, the Eurasian otter, and the European pine marten.

There are several characteristics that are common to all members of the family *Mustelidae*. In general, male members are larger than females. Their bodies are usually long and thin, although wolverines and badgers are stockier. Most species have short ears and limbs. Mustelids are diurnal or nocturnal and live in crevices and burrows. Many have the ability to sit up to look around. Mustelids have anal glands that secrete in a defensive measure. As carnivores, mustelids prey on other animals. Some can feed on plants, and some are omnivorous. They have keen eyesight and hearing, but generally hunt by smell.

Mustelids are mostly solitary except for Eurasian badgers, sea otters (*Enhydra lutris*) and some northern river otters (*Lutra canadensis*). In the more solitary species, association between males and females during the mating season is brief. Mating occurs mostly in the spring, and in many species, ovulation is induced during copulation. Delayed implantation of the fertilized egg occurs in many mustelids. Females raise the young alone. Only the least weasel produces two litters a year; other species produce them annually. In most mustelids, the young become sexually mature at about 10 months of age.

6.13. Evaluating Mustelid Species for Sedation and Anesthesia

Given their ubiquitous nature, anesthetizing mustelid species is often necessary in wildlife veterinary medicine. Certain species, such as ferrets, *Mustela nigripes* (black-footed ferret) are also popular pets, and many minks, including *Mustela vison* (new world mink) and *Mustela lutreola* (old world mink), are raised worldwide for their fur. Research and wildlife management also account for the necessity of the chemical immobilization of individual mustelids. Special challenges exist when anesthetizing smaller mammals, such as mustelids. First off, the risk of anesthesia-related death in small mammals is higher than in larger animals, even companion animals such as dogs and cats. The attendant stress can bring about tachycardia, tachypnea, hypertension, and other adverse effects, some of which can lead to death if too severe. Further, many smaller mammals cannot be intubated during general anesthesia, including most mustelid species.

Chemical restraint and anesthesia of the Mustelids (Wallach and Boever, 1983)

Animals	**Description of method**
Mink (5 lbs)	Ketamine HCl, intramuscular at 10-20 mg/kg + Xylazine HCl, intramuscular at 2 mg/kg.
	Ketamine HCl, intramuscular at 20-40 mg/kg to 100 mg/kg of body weight for major surgery.
	Tilazol (CI-144), intramuscular at 1.5-10 mg/kg of body weight
	Halothane, methoxyflurane, ether-ether jar or plastic bag.
	Reserpine, oral at 0.036-0.05 mg/animal.
	0.2 ml propriopromazine followed by 0.2 ml (1.2 mg) of pentobarbital for anesthesia of 5-6 hours.
	Pentobarbital, intraperitoneal at 40 mg/kg of body weight.
	Phencyclidine-promazine, intramuscular at 1-2 mg/kg.
Skunk	Halothane, methoxyflurane, ether-ether jar, plastic bag or anesthetic machine.
	Pentobarbital, intraperitoneal at 15-30 mg/kg.

	Pentobarbital, intrathoracic at 8 mg/kg of body weight for rapid induction.
	Phencyclidine-promazine at 1-2 mg/kg intramuscular.
	Ketamine HCl, intramuscular at 10-20 mg/kg + Xylazine HCl, intramuscular at 2 mg/kg of body weight.
	Ketamine HCl, intramuscular at 20-40 mg/kg of body weight.
	Tilazol (I-744), at 1.5-10 mg/kg of body weight.
Ferret	Halothane methoxyflurane, ether-ether jar, plastic bag or anesthetic machine.
	Pentobarbital sodium, intraperitoneal at 0.45g/500 g of body weight. It is having low margin of safety.
	Phencyclidine-promazine, intramuscular at 1.2 mg/kg of body weight.
Badger	Ketamine HCl, intramuscular at 10-20 mg/kg + Xylazine HCl intramuscular
	at 20-40 mg/kg of body weight.
Otter	Ketamine HCl, intramuscular at 20-40 mg/kg of body weight.
Grison	Tilazol CI-744, intramuscular at 1.5-10 mg/kg of body weight.
Turyn	Halothane, methoxyflurane, with gas anesthesia.
Weasel	Phencyclidine-promazine, intramuscular at 1-2 mg/kg of body weight.

At higher doses, some agents can be used to provide light surgical anesthesia (Green, 1978) in ferrets. Alphaxalone (12-15 mg/kg, intramuscular) produces its maximum effect in 10-12 minutes, and this effect lasts for 15-30 minutes. Full recovery takes 60-90 minutes.

Ketamine HCl (20-30 mg/kg intramuscular) produces a deep cataleptic stupor within 10 minutes of administration and lasts for 40-60 minutes. Muscular relaxation can be improved by the concurrent administration of xylazine HCl (1.0 mg/kg of body weight, intramuscularly).

Fentanyl citrate fluanisone (0.5 ml/kg, intramuscular) produces profound neuroleptanalgesia in 6-8 minutes, which lasts about 15-30 minutes. Muscular relaxation can be improved by the concurrent administration of xylazine HCl (1.0 mg/kg, intramuscular) or diazepam (2.0 mg/kg, intramuscular). Pentobarbitone sodium, given at 36 mg/kg, intraperitoneally, resulted in light surgical anesthesia in 6-10 minutes, which lasted for 50-120 minutes. However, full recovery is accompanied by muscular twitching, shivering, and limb paddling and may take several hours.

Fentanyl was used to produce neuroleptanalgesia in sea otter (Thomas *et al.*, 1981). It is administered intramuscularly at 0.05-0.11 mg/kg of body weight. In combination with azaperone at 0.11-0.45 mg/kg, fentanyl is administered at 0.05 mg/kg. When combined with azaperone at a dosage of 0.20 mg/kg, it met most of the requirements for a safe, short-acting intramuscular immobilization agent.

Five anesthetic agents, viz., C1744, etorphine, ketamine, and fentanyl were tested to establish dosage and safe, effective, short-acting anesthetics for use in sea otter (Williams and Kocher, 1978).

Etorphine: Dose is 0.75 mg/adult otter, used in conjugation with diazepam 1.25 mg/adult otter.

C1744: The induction is within 30 seconds to 6 minutes at 93 mg/kg body weight.The recovery time is 360 minutes.

Ketamine: It is contraindicated in debilitated otter as the thermoregulation center is affected.

Halothane: It is a safe and rapid immobilizer.

6.14. Immobilization of Bear

The bear belongs to the family Ursidae.
It includes the spectacled bear (*Tremarctos ornatus*), also known as the South American bear; the Asian black bear (*Ursus thibetanus*), also known as the Indian black bear; the American black bear (*Ursus americanus*), also known as the black bear; the polar bear (*Ursus maritimus*); the sun bear (*Helarctos malayanus*); and the sloth bear (*Melursus ursinus*), also known as the Indian bear.

The α-chloralose has been used as bait to capture American black bears (*Ursus americanus*). Chloralose is mixed with honey. The effective dose is 1gram per 3-8 lb of body weight. Recovery occurred in 8-10 hours. Ketamine HCl and xylazine HCl, combined in a ratio of approximately 2:1, produce good anesthesia in captive and wild black bears. The optimal dose range is 4.5-9.0 mg/kg of ketamine and 2.0-4.5 mg/kg of xylazine (Addition and Kulenusky, 1979). Free-ranging polar bears (*Ursus maritimus*) were immobilized using a concentrated solution of 200 mg of ketamine HCl and 200 mg of xylazine HCl/mL. A mean dosage of 6.8 mg of each drug/kg body weight was successful in immobilizing polar bears (Lee, 1981). 2.8 mg of each drug/kg body weight was effective for cubs. It has a wide safety margin, a lack of undesirable effects, no excess salivation, sufficient muscle relaxation, and no convulsions.

Wallah *et al.* (1967) have used etorphine HCl for the immobilization of bears. Patenaude (1979) evaluated fentanyl citrate, etorphine HCl, and Nalaxone HCl in captive polar bears. Nalaxone HCl does not cause papillary constriction, respiratory depression, and psychomimetic effects. Prime choice as antagonist at 2 mg/10 mg of fentanyl or 0.5 mg of etorpine administered by dart syringes. Smooth induction, greater degree of muscle relaxation are achieved with the use of fentanyl, either alone or in combination with etorpine. The immobilization of animals by fentanyl lasted 75 minutes. Animals recovered without any signs of discomfort. The effect of nalaxone took place within 10 minutes. The action first noticed was deep respiration and complete expiration, rising head and walking out smoothly.

Dosages of Etorphine and Diprenorphine (Alpord *et al.*, 1974)

	Etorphine at mg/4-5 kg of body weight.	**Diprenorpine at mg/4-5 kg of body weight.**
Black bear	0.5 (0.05 – 4.17)	1.0 (0.10 – 8.34)
Grizzy bear	0.5 (0.05 – 4.17)	1.0 (0.10 – 8.34)
Polar bear	0.5 (0.05 – 4.17)	1.0 (0.10 – 8.34)

Polar bears were immobilized using carfentanil at 1.0-38.0 μg/kg of body weight (Heigh *et al.,* 1983). Induction was rapid, within 5.0 minutes. The bears showed good muscle relaxation. The respiratory rate was depressed. The recurrence of narcotic effects, so-called recycling, was seen in some bears. The mean arousal after the administration of narcotic antagonists was longer than 5 minutes. Smaller bears can be anesthetized with intravenous thiopental, thiamylal, or pentobarbital sodium given at 13.5 mg/kg intravenously. Ketamine immobilization is reversed by Yohimbine (Page, 1986).

Useful Immobilization Drugs

Chemical immobilization of the species discussed here can be effectively and safely accomplished with the following drugs:

Agonist drugs	**Antagonist drugs**
butorphanol	atipamezole
detomidine	flumazenil
isoflurane	naltrexone
ketamine	naloxone
medetomidine	yohimine
midazolam	
propofol	
tiletamine/zolazepam (Telazol [TZ])	
xylazine	

7

Chemical Restraint of Insectivores

7.1. Introduction

Chemical restraint of insectivores is a technique used by veterinarians and wildlife biologists to immobilize small mammals that primarily eat insects, such as shrews, moles, and hedgehogs. It is essential for performing examinations, administering medications, and conducting research on these animals.

There are several factors to consider when chemically restraining insectivores:

- **Species:** Different species of insectivores may have varying degrees of sensitivity to different drugs.
- **Age and health:** Younger and healthier animals generally tolerate chemical restraint better than older or sick animals.
- **Temperature:** Insectivores are sensitive to temperature regulation, so it's crucial to maintain a stable body temperature during restraint.

Here are some of the common medications used for chemical restraint of insectaries:

- **Inhalant anaesthetics:** Isoflurane is a commonly used inhalant anaesthetic that provides rapid induction and recovery.

Isoflurane

- **Injectable anaesthetics:** Telazol (a combination of tiletamine and zolazepam) and ketamine are injectable anaesthetics that can be used alone or in combination.

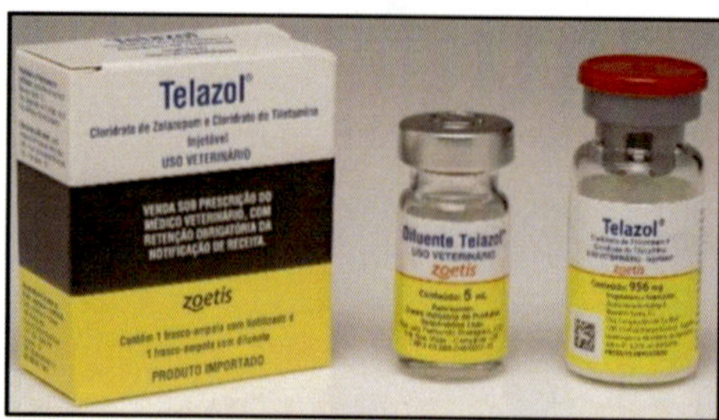

Telazol

- **Sedatives:** Midazolam and dexmedetomidine can be used to provide sedation for minor procedures.

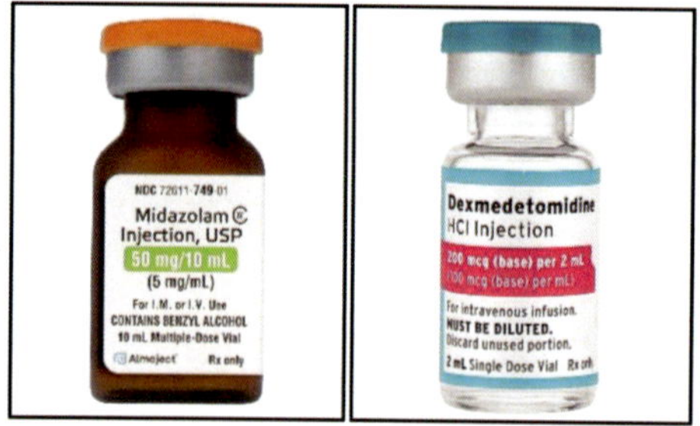

Midazolam and dexmedetomidine

The specific drugs and dosages used will vary depending on the individual animal and the procedure being performed. It is important to consult with a veterinarian or wildlife biologist experienced in chemical restraint of insectivores before attempting to restrain an animal yourself.Here are some of the reasons why chemical restraint might be necessary for an insectivore:

- To perform a physical examination
- To treat an injury or illness
- To collect samples for diagnostic testing
- To fit a tracking device
- To release a rehabilitated animal

Here are some additional points to remember:

- Chemical restraint should only be performed by trained professionals.
- Animals should be monitored closely during and after chemical restraint.
- Appropriate reversal agents should be readily available in case of overdose or other complications.

Chemical restraint should only be used when absolutely necessary, and all other options should be exhausted first. It is important to use the correct dosage of medication and to monitor the animal closely during and after the procedure. By following these guidelines, chemical restraint can be a safe and effective way to immobilize insectivores for necessary procedures.

7.2. Classification

The order Insectivora is actually an outdated classification used for small mammals that primarily feed on insects. While the term "insectivore" is still used informally to describe these animals, Insectivora itself is no longer considered a valid taxonomic order. It includes tailless tenrec (*Tenrec ecaudatus*), lesser hedgehog tenrec (*Echinops telfairi*), Eurasian hedgehog (*Erinaceus europaeus*), long-tailed shrew or rock shrew (*Sorex dispar*), Western American mole (*Scapanus townsendii*), Eastern mole (*Scalopus aquaticus*), Star-nosed mole (*Condylura cristata*).

Modern genetic studies have shown that the original Insectivora grouping was not a true evolutionary clade, meaning the animals classified within it weren't all descended from a single common ancestor. As a result, the insectivores have been reclassified into several different orders, including:

- **Soricomorpha:** This order includes shrews, moles, and solenodons.

Solenodon mammal

- **Erinaceomorpha:** This order includes hedgehogs and gymnures.

Hedgehog mammal

- **Afrosoricida:** This order includes golden moles, tenrecs, and otter shrews.

Golden mole mammal

- **Eulipotyphla:** This is a grandorder that encompasses all of the above orders, as well as several other families of insectivorous mammals.

7.3. Restraint and Immobilization of Insectivores

All insectivores are potentially pugnacious and should be handled with caution. Chemical restraint is necessary for most procedures involving insectivores.

Ketamine HCl and xylazine HCl combinations have been used with good success at a dose rate of 1.0-2.0 mg of ketamine HCl with 0.2 mg of xylazine HCl per gram of body weight. Induction time is 5-10 minutes and can be easily handled to facilitate extensive physical examinations and minor surgical procedures. The duration of anesthesia was 10-20 minutes.

The combination of phencyclidine and promazine at a rate of 0.1 mg/100 g of each, has also been used successfully. Pentobarbitone sodium, at a rate of 20 mg/500 g of body weight, has been used to anesthetize hedgehogs. This dosage provides deep anesthesia for up to 1.5 hours. Ether or halothane in a bell jar may also be used to anesthetize insectivores.

7.4. Restraint of Hyraxes

Hyraxes, also called dassies or coney, are small, thickset, herbivorous mammals in the order Hyracoidea. Hyraxes are well-furred, rotund animals with short tails. They are the only living members of the order Hyracoidea. It includes Syrian hyrax (Procavia capensis syriaca) and African rock hyrax (Procavia capensis). They are superficially similar to pikas and marmots, but are more closely related to elephants and sea cows.

Rock Hyrax

The dose for ketamine HCl is 15-20 mg/kg body weight, administered intramuscularly for analgesia and restraint. For anaesthesia, the dose is 25-40 mg/kg body weight, intramuscularly. Proper care should be taken to provide proper airway passage. Ketamine HCl and xylazine HCl at 10 mg/kg and 2 mg/kg body weight, respectively, can be administered intramuscularly. Phencyclidine and promazine, at 0.3-1.0 mg/kg body weight, can be administered for anesthesia.

7.5. Restraint of Edentates

Edentates, much like insectivores, require specialized restraint techniques due to their unique physiology and temperament. Is a mammal of an order distinguished by the lack of incisor and canine teeth, including the anteaters, sloths, and armadillos, all of which are native to Central and South America. Restraint should only be attempted by veterinarians or experienced wildlife biologists familiar with edentate anatomy and behaviour. Species, age, health status, and procedure all influence the chosen restraint method. The reasons for restraint are similar to those for insectivores, procedures like examinations, treatment, sample collection, tagging, or rehabilitation releases might necessitate restraint.

7.5.1.Restraint Methods for Edentates

- ***Inhalant Anaesthetics:*** Isoflurane is a popular choice for short procedures due to its fast-acting properties.
- ***Injectable Anaesthetics:***Telazol or ketamine are often used for injectables due to their effectiveness in edentates.
- ***Sedatives:*** Midazolam or diazepam can provide short-term sedation for minor examinations.

7.5.2. Specifics for Edentates

- ***Anteaters:*** Their long claws and powerful forelimbs necessitate extra caution during restraint. Induction boxes or specialized restraint cages might be used for initial capture.
- ***Armadillos:*** Their bony armor poses a challenge. Chemical restraint is often preferred over physical restraint to minimize injuries.
- ***Sloths:*** Their slow metabolism can affect how they metabolize drugs. Careful monitoring is crucial post-restraint.

7.5.3. Important Considerations

- ***Minimize Restraint:*** Chemical restraint should be a last resort. Alternative methods like food deprivation or short-term physical restraint (for examinations) should be explored first.
- ***Correct Dosage:*** Using the appropriate amount of medication is vital for animal safety.
- ***Close Monitoring:*** Animals should be closely monitored throughout the procedure and recovery to ensure their well-being.

Order Edentata includes nine banded armadillo (*Dasgpus novemcinctus*), giant anteaters (*Myrmecophaga tridactyla*), Tamanduas (*Tamandua tetradactyla*), two-toed sloths (Choloepus species), pangolins or scaly anteaters (three genera namely Manis, Phataginus, and Smutsia), aardvark (*Orycteropus afer*)

Ketamine HCl and xylazine HCl at 10 mg/kg and 2 mg/kg of body weight, respectively, can be administered intramuscularly. Ketamine HCl alone may be given intramuscularly at 1-20 mg/kg of body weight. It produces effects ranging from tranquilization to surgical anesthesia. Tiletamine HCl and zolazepam can be administered at a rate of 1.9-6.0 mg/kg of body weight. Droperidal and fentanyl combination (Innovar Vet) has been extensively used in Armadillos at 0.20-0.25 ml/kg of body weight administered intramuscularly. Phencyclidine HCl and promazine HCl, at a rate of 0.5-0.8 mg/kg of body weight, can be administered to safely immobilize aardvark, pangolin, sloth, giant anteater, and tamandua. Pentobarbital at a rate of 25-35 mg/kg can be used intraperitoneally for surgical anesthesia. Atropine sulphate is administered to control salivation, which is a potential hazard. It is given at a rate of 0.04 mg/kg body weight, subcutaneously, and an endotracheal tube should be inserted. Gaseous anesthesia, halothane, and methoxyflurane may be used to maintain anesthesia. Local anesthesia may be used to provide regional analgesia.

7.6. Anaesthesia and Restraint of Bats

7.6.1. Classification

Bats belong to the Order Chiroptera. It includes fruit bats (*Pleropus* species), mouse-tailed bats (*Rhinopoma* species), sheath-tailed bats (*Taphozous* species), fisherman bats (*Noctillo* species), slit-faced bats (*Nycteris* species), false vampire bats (*Megaderma* species), horseshoe bats (*Rhinolophus* species), leaf-nosed bats (*Hipposideros* species), vampire bats (*Desmodus* species), smoked bats (*Turipterus* species), disk-winged bats (*Thyroptera* species), little brown bats (*Myotis* species), New Zealand short-tailed bats (*Mystacina* species), and mastitt bats (*Promops* species).

Bats, with their unique biology and nocturnal habits, require specific approaches to anaesthesia and restraint. Here's a breakdown of the key points:

7.6.2. Why Anesthetize or Restrain Bats?

- **Minimizing Stress:** Procedures like sample collection, examinations, or treatment can be stressful for bats. Anaesthesia or restraint helps reduce this stress for both the animal and the handler.
- **Safety:** Bats can bite or scratch, potentially transmitting zoonotic diseases. Restraint minimizes this risk for researchers and veterinarians.
- **Effective Procedures:** Certain procedures require a still bat for accuracy and safety. Anaesthesia ensures this stillness.

7.6.3. Anaesthetic Options for Bats

- **Inhalant Anaesthetics:** Isoflurane is a common choice due to its rapid induction and recovery times and minimal metabolism in the bat's body.
- **Injectable Anaesthetics:** Combinations like ketamine and xylazine or dexmedetomidine and ketamine are effective for short-term procedures.

7.6.4. Restraint Techniques for Bats

- **Physical Restraint:** For brief examinations, a gloved hand can gently restrain the bat, with a finger placed over the wings to prevent flight.
- **Restraint Bags:** Fabric pouches with openings for the head can be used for short-term restraint during procedures.

Important Considerations

- **Species Differences:** The best method depends on the bat species, size, and health. Consult a veterinarian experienced with bats for specific protocols.
- **Minimal Handling:** Handle the bats for the shortest time possible to minimize stress.

- **Temperature Regulation:** Maintain a warm environment during restraint to prevent hypothermia in bats.

Bats are difficult to handle and anesthetize due to their thermoregulatory system. Ketamine HCl is used in little brown bats at a rate of 40 mg/kg for immobilization and 120 mg/kg to produce anesthesia lasting 20-30 minutes. Ketamine HCl and xylazine HCl combinations at10-20 mg/kg and 2 mg/kg, respectively, are effective. Pentobarbitone at 30-50 mg/kg given intraperitoneally is effective. It is necessary to maintain a rectal temperature of 37-40°C during induction and anesthesia. If rectal temperature begins to drop, additional heat should be provided by using a hot water blanket, incubator, heat lamp, or any other heat source. Inhalant anesthetics such as halothane, methoxyflurane, and ether can be used in bats by utilizing a bell jar and then maintaining the animal with gas delivered through a cone.

7.7. Immobilization of Marsupials

7.7.1. Classification

Marsupials, like kangaroos, koalas, and wombats, present unique challenges when it comes to immobilization. Their physiology and behaviour require careful consideration to ensure their safety and well-being during procedures.

The Order *Marsupialia* includes wooly opossum (*Caluromys* species), mouse opossums (*Marmosa* species), Virginia opossums (*Didelphis virginiana*), bandicoots (*Perameles* species), long-nosed bandicoots (*Perameles nasuta*), brush-tailed opossums (*Trichosurus vulpecula*), koalas (*Phascolarctos cinereus*), rock wallabies (*Petrogale penicillate*), yellow-footed rock-wallaby, formely known as ring-tailed rock-wallabies (*Petrogale xanthopus*), black-striped wallaby, also known as scrub wallaby or eastern brush wallaby (*Notamacropus dorsalis*), and Kangaroos (largest of marsupial) (*Macropus* species).

The Order *Monotremata* includes duck-billed platypuses (*Ornithorhynchus anatinus*), spiny anteaters (*Tachyglossus acceleatus*), and long-beaked newguinea echidnas (*Zaglossus bruijnii*).

7.7.2. Methods of Immobilization for Marsupials

- **Physical Restraint:** For short examinations or pouch handling, physical restraint using a towel or a specially designed pouch restraint may be used. This should only be done by experienced personnel.
- **Chemical Restraint:** For procedures requiring muscle relaxation or sedation, injectable or inhalant anesthetics are used.

- **Injectable Anesthetics:** Telazol (a combination of tiletamine and zolazepam) and ketamine are commonly used options due to their effectiveness in marsupials.
- **Inhalant Anesthetics:** Isoflurane is a popular choice for short procedures because of its fast-acting properties.

7.7.3. Considerations for Choosing a Method

- **Species:** Different marsupials have varying sensitivities to drugs. A veterinarian who is familiar with the specific species should determine the safest and most effective method.
- **Age and Health:** Age, weight, and any underlying health conditions will influence the chosen immobilization technique.
- **Procedure:** The complexity and duration of the procedure will determine the level of immobilization needed.

Additional Considerations

- **Pouch Young:** Special care is required for pouch young during the mother's immobilization. They may need to be removed and kept warm in a pouch surrogate.
- **Monitoring:** Close monitoring of vital signs like heart rate and respiration is crucial throughout the immobilization period.
- **Recovery:** A quiet recovery area with appropriate temperature control is essential for the marsupial to wake up safely.

Oral chlordiazepoxide has been used successfully as a tranquilizer for kangaroos at 11 mg/kg body weight, orally. The induction occurs in 60-90 minutes. Diazepam in Virginia opossum is 20-110 mg/kg in the food. The induction occurs in 2-3 hours. Pentobarbital sodium at 25 mg/kg can be used for kangaroos. Opossum at 36 mg/kg for slight duration procedures in kangaroos and wallabies. Thiamylal sodium, or methohexital sodium, can be used at 25 mg/kg.

The drugs of choice for immobilizing marsupials are ketamine HCl (10 mg/kg body weight) and xylazine HCl (2 mg/kg body weight) given intramuscularly, offer a high safety margin and good immobilization. Ketamine HCl in Virginia opossums is 20-25 mg/kg body weight. The fentanyl and droperidal combination (Innovar vet) is 0.75-10 mg/kg body weight. For marsupials,CI-744 (Tilazol) is used at 5 mg/kg body weight. Pentobarbital and chloral hydrate (Equithesin) at 2 ml/kg body weight. Phencyclidine HCl and promazine HCl can be used at 0.7-10 mg/kg of body weight. In gas anesthesia halothane or methoxyflurane can be used.

Small kangaroos were anesthetized using sodium thiopental intravenously for induction. Endotracheal intubation was difficult because of it prevented adequate visualization of the larynx. Intubated by using a laryngoscope and a straight wire stylet inserted into the lumen of the tracheal tube to facilitate endotracheal intubation (Richardson and Cullen, 1981).

7.8. Immobilization of Wild Swine and Peccaries

7.8.1. Classification

Wild swine and peccaries belong to the Order Artiodactyla. Order Artiodactyla is further divided into family Suidae and family Tayassuidae (Peccaries). The family Suidae includes European wild boar (*Sus scrofa*), bush pigs (*Potamochoerusporcus*), warthog (*Phacochoerus aethiopicus*). The family Tayassuidae (Peccaries) includes collared peccaries (*Tayassu tajacu*) and white-lipped peccaries (*Tayassu pecari*).

7.8.2. Chemical Restraint: The Preferred Method

Wild swine (including feral hogs) and peccaries are powerful and potentially dangerous animals. Their immobilization requires specific techniques to ensure safety for both the animal and the handler.

Due to the risks involved in physically restraining these wild animals, chemical restraint is the preferred method for immobilization.

- **Common Drugs:** A combination of Telazol (tiletamine and zolazepam) and xylazine HCl is a common and effective choice for both wild swine and peccaries.
- **Dosage:** The appropriate dosage depends on the animal's weight and species. A veterinarian familiar with large mammals should determine the specific amount.

Delivery Methods

- **Remote Injection:** Delivering the drugs via darts fired from a distance is the safest method. It minimizes stress on the animal and reduces the risk of injury to the handler.
- **Trapping and Injection:** In some cases, the animal might be lured into a trap where it can be safely injected with the immobilizing drugs.

Important Considerations

- **Species Differences:** While telazol-xylazine works well for both, slight variations in dosage or alternative medications might be needed for specific peccary species. Consulting a wildlife veterinarian is crucial.

- **Experience Matters:** Immobilization of wild swine and peccaries should only be attempted by trained professionals with experience in handling large wildlife.
- **Rapid Recovery:** These drugs typically provide short-acting anesthesia, allowing the animal to recover relatively quickly after the procedure.
- **Monitoring:** Close monitoring of vital signs like respiration and heart rate is essential during immobilization and recovery.

Alternatives to immobilization (when possible):

- **Fencing and Exclusion Techniques:** For situations where population control is the goal, exclusion fencing or trapping methods might be preferable to avoid the need for immobilization altogether.

Only intramuscular drugs are practical in these families. Projectile syringes, blow guns, and pole syringes are used to inject the drug. Ketamine HCl is administered at 10-30 mg/kg intramuscularly. Lower dosages produce tranquilization and immobilization, and higher dosages produce short-term surgical anesthesia. Induction occurs within 2.2-2.8 minutes, and the maximum level is reached at 8.5 minutes. The duration is 10-20 minutes. Atropine sulphate is indicated at a rate of 0.044 mg/kg body weight to control salivation. Phencyclidine HCl and promazine HCl at a rate of 0.7-30 mg/kg each given intramuscularly produce a dose-related effect ranging from immobilization to surgical anesthesia in 12-20 minutes.

Etorphine (M-99) is used at a rate of 1.0-30 mg of total dose for adult swine and 1.0-2.0 mg for peccaries. The ear veins, or posterior vena cava, are used to inject the antagonist. An endotracheal intubation can be done only after they have been immobilized. Pharynx has been sprayed with 2% lidocaine. Anesthesia can be maintained with inhalant anesthetics. Extra caution should be taken when immobilizing any swine who is extremely pronc to hyperthermia, especially at higher anesthetic doses.

Arora (1996) used Immoblin to immobilized escaped wild pigs(Sus scrola cristatus). The drug Immoblin contains Etorpine HCl 2.45 mg/ml and Acepromazine 10 mg/ml. A total dose of 0.8ml to 1.2ml for adults and 0.25ml to 0.5ml per young animal (aged about 1 year) was used. The induction time ranged between 7-15 minutes.

Doses of intravenous general anesthesia for exotic swine (Wallach and Beever, 1983)

Drug	Body weight (kg)	mg/kg	Duration of effect (min)	Recovery time following surgery (min)
Thiopental sodium	5-20	11	4-70	30
Thiopental sodium	20-45	10	-do-	-do-
Thiopental sodium	45-90	9	-do-	-do-
Thiopental sodium	90-135	8	-do-	-do-
Thiomylal sodium	All weights	17	10-20	30
Pentobarbital sodium	All weights	18-28	45-60	Extended (4-6 hours)

Additional Resources

- Immobilization of Collared Peccaries (Tayassu tajacu) and Feral Hogs (Sus scrofa) with Telazol® and Xylazine [link to research paper on immobilization of feral hogs and peccaries] (This research paper explores the use of Telazol-xylazine combination for immobilizing both wild swine and peccaries)
- Standard Operating Procedure for the Study of Bats in the Field https://www.nps.gov/subjects/bats/index.htm (This guide provides a good overview of restraint methods for field studies)
- Comparison of two injectable anaesthetic protocols in Egyptian fruit bats (Rousettus aegyptiacus) undergoing gonadectomy [This research paper discusses anesthetic options for bats undergoing procedures]
- Capture and Immobilization of Free-Ranging Edentates https://www.scribd.com/document/64291435/History-of-Entomology (This PDF resource provides detailed information on edentate restraint techniques)

8

Chemical Restraint of Wild Ruminants

8.1. Classification

The ruminants belong to the Order Artiodactyla. The families in order of Artiodactyla include Camelidae, Tragulidae, Cervidae, Giraffidae, Antliocapridae, and Bovidae.

The family Camelidae includesGuanaco (*Lama guanicoe*), Llama (*Lama glama*), Alpaca or South American camelid mammal (*Lama pacos*), Dromedary camel (*Camelus dromedarius*), and Bactrian or two-humped camels (*Camelus bactrianus*).

The family Tragulidae includes Chevrotains originate from Africa (*Hyumoschusaquaticus*) and Chevrotains that originate from Asia (*Tragulus species*).

The family CervidaeincludesWhite tailed deer(*Odococlius virginianus*), Follow deer (*Dama dama*), Roe-deer (*Capreolus capreolus*), Axis deer (*Axis axis*), Reindeer (*Rangiter tarandus*), Musk deer (*Moschus species*) and Chinese water deer (Hydropotes species).

The family Giraffidae includes Giraffe (*Giraffa camelopardis*) and Okapi (*Okapia johnstoni*).

The family Antliocapridaeinclude Prong horn antelope (*Antilocapra Americana*).

The family Bovidae includes Sitatunga or marshbuck (*Tragelaphusspekii*), Greater kudu (*Tragelaphus strepsiceros*), bush buck (*Tragelaphus scriptus*), Eland (*Tragelaphus Raurotragus oryx*), Nilagi (*Bostaphustragocamelus*), Water buffalo (*Bubalus bubalis*), Yak (*Bos grunniens*), American bison (*Bison bison*), Wisents (*Bison bonasus*), Duinkers (*Cephalaphus species*), Water bucks and Kobs (*Kobus species*) Common water buck (*Kobus ellipsiprymnus*), Sable antelope (*Hipppotragusniger*), Roan (*Hipppotragus equines*), Oryx and gems bok (*Oryx species*), Addax (*Addax nasomaculatus*), Harle beests (*Alcelaphus*species), Wild beests (*Connochaetes*species), Indian black buck (*Antilope cervicapra*), Musk ox (*Ovibosmoschatus*), Wild goats (*Capra species*), Spanish ibex (*Capra pyrenaica*), Siberian ibex (*Capra ibex sibirica*), Retan wild goat (*Capra aegaguescretica*), Aoudad (*Ammotragus *lervia*),Moutlon (*Ovis musimon*), Bighorn North American wild sheep (*Ovis canadensis*), and Dall sheep (*Ovis dalli*).

8.2. Intramuscular Agents for Immobilizing Wild Ruminants

Ketamine: A good fast-acting drug for immobilization. Short recovery period and poor muscle relaxation when given alone. Could be used for a simple examination or for simple immobilization.

Xylazine: Results were variable. The drug was usually fast-acting and gave good muscle relaxation. Animals were immobilized until they were stressed or approached. Got up when provoked. Recovery periods were long.

Etorphine:Thetime to the onset of immobilization was a little longer than ketamine and xylazine alone. No muscle relaxation, excellent recovery following administration of antagonists (M50-50 or M285). An excellent drug for simple immobilization.

Ketamine with Etorphine:The time to onset of immobilization was probably shortened, but no real advantage was gained by the combination. Since no muscle relaxation resulted.

Xylazine with Etorphine:The time of onset of immobilization was shorter than the Etorphine alone. Excellent muscle relaxation for almost any surgical procedure. There are no problems with reversal of effect using antagonists (M-50-50, M285) if antagonistsarenot given within 90 minute after injection of the anesthetic. This combination offers the advantages of excellent muscle relaxation and the opportunity for rapid reversal of drug action.

Xylazine with Ketamine: Provided quick immobilizing action and good muscle relaxation. Recovery is more rapid than when anesthesia is induced with xylazine HCl alone, because ketamine metabolized, faster than xylazine. Xylazine with Etorphine and Xylazine with Ketamine are good combinations for inducing anesthesia in ruminants.

8.3.Immobilization of Camels

There are two main methods for immobilizing camels: chemical immobilization and physical restraint.

8.3.1. Chemical immobilization: It is the most common method and involves using drugs to sedate or anesthetize the animal. This is typically done by injecting the drugs through a dart fired from a distance using a specialized rifle. Chemical immobilization should only be performed by a trained veterinarian or other qualified professional, as there are risks associated with the drugs and the immobilization process itself.

8.3.2. Physical restraint: It is less common but may be necessary in some situations. This can involve using ropes or hobbles to restrain the animal's legs. Physical restraint should be used with caution, as it can be stressful for the camel and can lead to injuries.

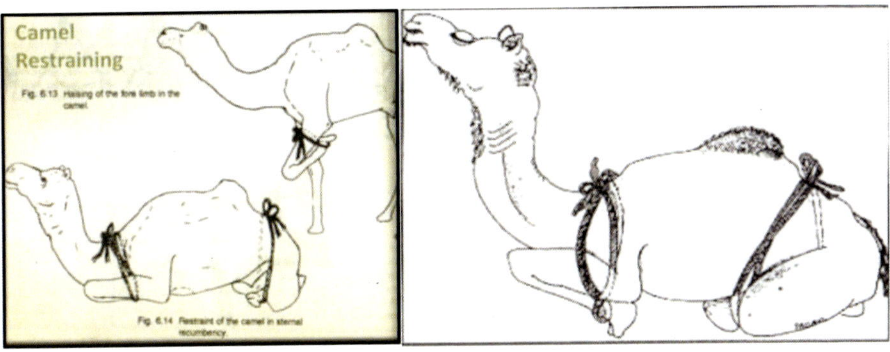

Physical restraint

Here are some of the factors to consider when choosing a method of immobilization:

- The temperament of acamel.
- The reason for immobilization.
- The availability of a trained veterinarian or other qualified professional.
- It is important to remember that all methods of immobilization carry some degree of risk. It is important to weigh the risks and benefits before choosing a method.

Xylazine HCl has been used in most domestic animals (Bauditz, 1972). Camels are sensitive to xylazine as a sedative. Xylazine has been used at 0.4 mg/kg of body weight. The onset of weak time, down time, and time until recovery was 8.6±1.1, 10.5±0.6, and 150±56.9 minutes, respectively. The analgesic effect remained for about 60-90 minutes (Peshin*et al.,* 1980). Xylazine induced sufficient muscle relaxation, i.e., relaxation of the neck, tail, and anus sphincters.

Weak time: time elapsed from administration of xylazine to the onset of ataxia-staggering gait, inability to balance on limbs, and abduction of limbs in an attempt to balance the body.

Down time: time elapsed from administration of xylazine HCl to the onset of sternal recumbency.

Recovery time: time elapsed from administration of xylazine HCl to voluntary standing of the animal.

Regurgitation of rumen contents and ruminal tympany were not observed in Camelidae or Bovidae. Recovery was not accompanied by straggling and excitement.

Bactrian camels were anesthetized using xylazine HCl at arate of 0.27-0.51 mg/kg of body weight given intramuscularly ,(Custer *et al.,* 1977). Camels anaesthetized have significantly lower values for RBC, Hb, and PCV and significantly higher blood glucose concentrations. There are no significant acid-base disturbances. Rapid arousal of the sedated camel after the intravenous administration of 0.05-0.13 mg/kgdoxapramHCl.

Species	**Sedation**	**Immobilization**
Bactrian or two humped camels*(Camelus bactrianus)* Dromedary camel (*Camelus dromedarius)*	0.1-0.5 mg/kg of body weight, intramuscularly	1-2 mg/kg of body weight, intramuscularly
Guanaco *(Lama guanicoe)*	0.3-1 mg/kg of body weight, intramuscularly	2 mg/kg of body weight, intramuscularly
Llama *(Llama glama)*	0.2-0.5 mg/kg of body weight, intramuscularly	1-2 mg/kg of body weight, intramuscularly

Chloral hydrate (10%) and magnesium sulfate (10%) solutions can be given intravenously to acamel. Another formula is chloral hydrate (2 oz) and$MgSO_4$ (2 oz)mixed in 16 oz of distilled water. Premedicated with chlorpromazine (1mg/kg body weight, intramuscular) to reduce the amount of chloral hydrate.

Chlorpromazine (0.5 mg/kg of body weight), 2.5g/150 kg of chloral hydrate intravenous administration, followed by an average of 2.33g of pentobarbital sodium given intravenous till effect. This mixture produced surgical anesthesia. For reversal, use doxapram HClor yohimbine in llamas. Local anesthetics may be used for laparotomy or rumenotomy (Riebord, 1986).

Guaifenesin (110 mg/kg, intravenous) and thiopental (4.4 mg/kg, intravenous) have been used for induction of anesthesia in a bull dromedary camel. For continuation of general anesthesia, guaifenesin (in 5% dextrose) plus thiopental (0.2%) in the same solution was given along with N_2O-O_2halothane anesthesia (Lumb and Jones, 1984).

8.4. Immobilization of Giraffe

There is one species of giraffe *(Giraffa camelopardalis)* with nine subspecies currently recognized. There are three subspecies which are the most wide-ranging and typically encountered by wildlife managers, usually in sub-Saharan nations such as Kenya and Namibia:

Giraffa camelopardalis tippelskirchi - Commonly called the Maasai giraffe, this subspecies is native to central and southern Kenya.

Giraffa camelopardalis reticulate - Called the Reticulated or Somali giraffe, this subspecies is found in northern parts of Kenya

Giraffa camelopardalis rothschildi - The Rothschild, Baringo or Ugandan giraffe is usually found in protected areas of Kenya including Mt. Elgon, Ruma and lake Nakuru National Parks, Mwea national reserve and several private conservancies/facilities

Giraffes, due to their unique anatomy and size, present a challenge when immobilization is needed. Immobilization of giraffes is a dangerous procedure that should only be attempted by a highly trained veterinarian or wildlife biologist with a team of experienced personnel.

Here's a breakdown of the process:

8.4.1. Methods of Immobilization

The most common method for immobilizing a giraffe is through chemical immobilization. This involves using specialized darts to deliver a combination of drugs that will sedate the animal. The specific drugs used will vary depending on the age, health, and size of the giraffe, but they typically include:

- **Opioids:** These drugs produce a state of sedation or unconsciousness. Examples include etorphine and carfentanil.
- **Tranquilizers:** These drugs help to calm the animal and reduce anxiety. Xylazine HCl is a common tranquilizer used in giraffe immobilization.
- **Reversal agents:** These drugs are used to reverse the effects of the immobilization drugs once the procedure is complete. An example is diprenorphine, which reverses the effects of etorphine.

Here's a general outline of the chemical immobilization process for a giraffe:

1. **Preparation:** The area where the immobilization will take place should be prepared beforehand to minimize stress on the animal. This may involve creating a safe enclosure or using crowd control to keep other giraffes at a distance. The veterinarian will also gather the necessary equipment and drugs.
2. **Darting:** The veterinarian will use a specialized dart rifle to deliver the immobilization drugs. The dart is typically fired at the giraffe's rump or shoulder.
3. **Induction:** Once the drug is injected, it will take some time for the giraffe to become sedated. The veterinarian will closely monitor the animal during this phase, which can last up to 15 minutes.
4. **Monitoring:** Once the giraffe is sedated, the veterinarian will carefully lower the animal to the ground if necessary and position it for the planned

procedure. Critical body functions like respiration and heart rate will be monitored throughout the immobilization period.

5. **Reversal:** At the conclusion of the procedure, an antagonist drug will be administered to reverse the effects of the immobilization drugs and allow the giraffe to recover.

8.4.2. Risks of Immobilization

Immobilization is a stressful procedure for giraffes, and there are a number of risks involved, including:

- **Respiratory depression:** Giraffes have a relatively low respiratory rate, and immobilization drugs can further depress their breathing.
- **Cardiovascular collapse:** The drugs used for immobilization can also cause a drop in blood pressure and heart rate.
- **Muscle damage:** If a giraffe falls or struggles during immobilization, it can sustain muscle damage.
- **Neck injury:** The long neck of a giraffe is particularly vulnerable to injury during immobilization.

8.4.3. Precautions to Minimize Risks

To minimize the risks of immobilization, a number of precautions must be taken. These include:

- **Pre-immobilization assessment:** A veterinarian will carefully examine the giraffe to assess its overall health and suitability for immobilization.
- **Proper dart placement:** The drugs must be injected into a muscle mass for rapid absorption.
- **Monitoring:** Once the giraffe is sedated, it must be continuously monitored by a team of experienced personnel. This monitoring will include checking the animal's heart rate, respiratory rate, and body temperature.
- **Positioning:** The giraffe's head and neck must be positioned in a way that minimizes the risk of regurgitation and aspiration.
- **Reversal of anesthesia:** Once the procedure is complete, the reversal agent will be administered to bring the giraffe out of sedation.

Here are some additional important points to consider regarding giraffe immobilization:

- **Fast and Efficient Immobilization:** The goal is to achieve rapid and effective immobilization to minimize the risks associated with the drugs and the stress on the animal.

- **Head and Neck Support:** The long neck of a giraffe is a potential danger during immobilization. The veterinarian will take steps to support the head and neck to prevent injuries.
- **Minimal Immobilization Time:** The period of immobilization should be kept as short as possible to reduce the risk of complications.

Immobilization in the giraffe is a challenge because of its unique anatomy and physiology. These present inherent problems during chemical restraint, including:

- The giraffe is a very large animal, with males weighing 790-1,400 Kg and females weighing 700-950 Kg. This limits effective physical control during the critical times of induction and recovery, and limits manipulation once the animal is down.
- If not controlled, the giraffe's long neck can act as a lever arm, creating a danger to the animal and capture personnel. A poorly-positioned neck can also lead to airway obstruction and/or cramping of neck muscles, which has led to fatalities in the field.
- The giraffe's long legs potentiate self-induced injury due to slipping during induction and recovery.
- Giraffes have an unfortunate tendency to regurgitate under the stress of capture and immobilization, which can lead to fatal aspiration pneumonia; further, the posterior position of the larynx in the pharynx hampers the draining of rumen and/or saliva. According to Kenya Wildlife Service, "[V]omiting can result from the increased intra-abdominal pressure occurring when the animal impacts the ground since the skin and muscles over the abdomen are very tense. A rumen bolus can on occasion be seen as it progresses up the neck in some giraffes receiving opioids just prior to or during recumbency."
- Prolonged induction and/or recovery can lead to hyperthermia, myopathy and secondary trauma. While this is true of nearly all exotic mammal species, this is more pronounced in the giraffe, since this species has elevated systolic blood pressure in order to maintain adequate perfusion in the brain.
- The giraffe's heart has high energy and oxygen requirements to maintain its high blood pressure. If breathing is impaired, even briefly, there is a pronounced danger for heart failure.
- The giraffe has a small respiratory tidal volume with a large dead space and relative small cardiac output during anesthesia. This results in a limited exercise tolerance and considerable resistance to air moving

through the long respiratory tract. Thus, while an immobilized giraffe may appear to be breathing, it still may not be getting sufficient oxygen, putting it at risk for heart failure.

- The giraffe has the thickest skin of all ungulates; thus, darting needles must be long (over 50mm) and of sufficient bore gauge to effectively penetrate the skin.
- In most cases, a giraffe will stand up with difficulty after immobilization, especially if a tranquilizer has been used in the immobilization formulation. This potentiates risk factors for injury when the animal is released. *(Kenya Wildlife Service, 2019)*
- While it may be self-evident at this point, the size of the giraffe features prominently in the success of any capture procedure, and smaller animals tend to have better a success rate than the larger adults, since the former are more easily immobilized and restrained (Bush *et al.*, 1976; Immobilization and Translocation Protocol for the Giraffe (Giraffa camelopardalis) in Kenya (2019).

Xylazine HCl alone is unsuitable for the restraint of giraffes. However, when it is combined with etorphine,satisfactory results are obtained. Doses for Masai and Reticulated Giraffes range from 20-70 mg/kg of xylazine plus 0.8-1.5 mg of etorphine. The dose of succinylcholine chloride for Masai Giraffe *(Giraffa camelopardis)* is 0.024 mg/lb and is rather critical. This differs considerably from the dose for Rothschild's Giraffe, which is 0.08 mg/lb.

Etorphine 2.5 mg has been used for surgical analgesia and restraint in an adult Reticulated Giraffe, along with 30 mg of acepromazine and 10 mg of scopolamine given intramuscularly. This produced recumbency but was supplemented with an additional 1 mg of etorphine to produce surgical analgesia. Nalorphine 250 mg may be given intramuscularly or intravenously to quickly counteract the etorphine.

Dosage of M-99 and additives for the chemical restraint in Giraffe (Wallach and Boever, 1983)

Body weight (kg)	M-99 Etorphine (mg)	Acepro - mazine (mg)	Scopo - lamine (mg)	Immobi - lization time (min.)	Duration min (min.)	Antagonist and Recovery time	
1500	2.5	30	10	6	60	N-500	4-5 min
2000	3.5	30	20	8	90	C7	4-8 min

Etorpine (M-99) dosage as per Wallach and Boever, 1983 in Giraffe.

The total dose for adults is 3-8.0 mg. The total dose for juveniles (6 months to 2 years) is 2.0-5.0 mg.

For the immobilization of giraffeArnemo and Kreeger (2018) recommends thiafentanil (20 mg for males, and 14 mg for females). This can be supplemented with 5 mg of thiafentanil if needed. For reversal, the authors recommend naltrexone (200 mg for males, and 140 mg for females). Alternatively, etorphine may be used at 16 mg for males, and 12 mg for females. This may be reversed with naltrexone at 10 mg for each mg of etorphine given.

Nielsen (1999) recommends etorphine in total dosages of 5 to 6 mg with 20 mg of xylazine, or carfentanil in total dosages of 8 mg with 10 mg of atropine and 100 mg of xylazine. The author also reports that in some regions of Africa, etorphine in total dosages of 8 to 10 mg has been used in adult giraffe.

Once the animal is down, the giraffe's neck should be extended to ensure the airway. The neck should also be supported by at least two capture personnel, with the head maintained above the rumen and the nose pointed down to facilitate drainage of any rumen or pharyngeal fluids.[3]

8.5. Immobilization of Antelope and Gazelle

Antelope and gazelles, being swift and often wary animals, require careful immobilization techniques to ensure their safety and the success of the procedure. Similar to other large animals, there are two main methods: chemical immobilization and physical restraint.

8.5.1. Chemical Immobilization

This is the preferred method for antelope and gazelles due to its efficiency and control. Here's a breakdown:

- **Drugs:** Veterinarians will choose a specific drug combination based on the species, size, and health of the animal. Factors like age, temperament, and the intended procedure also influence drug selection.
- **Delivery:** Drugs are typically delivered via dart fired from a distance using a specialized rifle. This minimizes stress on the animal and allows for intervention in less controlled environments.
- **Induction and Monitoring:** After the dart hits, it takes some time for the drug to take effect. A trained professional will closely monitor the animal during this induction phase, which can last around 10-15 minutes. Once sedated, the animal's vital signs, like respiration and heart rate, are monitored throughout the immobilization.
- **Reversal:** Once the procedure is complete, an antagonist drug is administered to reverse the effects of the immobilization drugs, allowing the animal to recover safely.

8.5.2. Physical Restraint

Physical restraint is less common for antelope and gazelles due to the risk of injury to both the animal and the handlers. It may only be used in specific situations where darting is not possible. Here's what to consider:

- **Limited Use:** Physical restraint should only be attempted as a last resort and only by experienced personnel.
- **Stressful for Animals:** This method can be very stressful for antelope and gazelles, increasing the risk of complications.
- **Potential Injuries:** The struggle to restrain the animal can lead to injuries for both the animal and the handlers.

Important Considerations

- **Training and Expertise:** Immobilization of any wild animal should only be attempted by a trained veterinarian or wildlife professional with the necessary experience and knowledge.
- **Minimize Risks:** The goal is to achieve rapid and effective immobilization with minimal stress on the animal and ensure a safe recovery.
- **Species Variations:** The specific immobilization techniques may vary depending on the antelope or gazelle species being dealt with.

Additional Tips

- Having a contingency plan in case of complications during immobilization is crucial.
- Minimizing human presence around the animal during the recovery phase can help reduce stress and promote a smooth awakening.
- Remember, immobilization of wild animals is a serious procedure and should only be undertaken when absolutely necessary.

Gazelles are famously fleet-footed members of the antelope family. Pronghorn antelope may be immobilized using capture equipment and 6.05 to 9.93 mg/100 lbs of succinylcholine chloride.

Xylazine HClat 1.0-3.0 mg/kg is effective for sedation and the capture of wild ruminants and antelope. The dosage for xylazine is given below (Bauditz, 1972).

Species	Dose for Sedation	Dose for Immobilization
Gazelle	1 mg/kg of body weight	2-4 mg/kg of body weight
Black buck	1 mg/kg of body weight	3 mg/kg of body weight
Spring buck	1 mg/kg of body weight	1-3 mg/kg of body weight
Impala	1 mg/kg of body weight	3 mg/kg of body weight

Capture myopathy is a fatal condition in captured antelopes. There havebeen reported deaths in Pronghorns *(Antilocapra americana).* The clinical signs include depression, stiffness, weakness, incoordination, recumbency, paralysis, and myoglobinuria. The postmortem finding indicates extensive damage to muscles and other organs as a result of capture procedures (Chalmers and Barrett, 1977).

Bighorn sheep: Etorphine at arate of 1.77 mg/4-5 kg body weight is initially used intramuscularly. After that,acepromazine is used to prolong the effect and facilitate handling. It is antagonized by diprenorphine at 3.54 mg/45 kg (total dosage: 0.38-26.6 mg) given intravenously (Harhoorn, 1966).

8.6. Immobilization of Family Bovidae

The Bovidae family encompasses a wide range of hoofed mammals, including antelope, gazelles, bison, wildebeest, cattle, sheep, and goats. Due to the diversity within the family, specific immobilization techniques can vary depending on the species. However, some general principles apply:

8.6.1. Chemical Immobilization

This is the preferred method for most Bovidae members due to its control and efficiency. A veterinarian will select a drug combination based on the species, size, health, age, and temperament of the animal, as well as the intended procedure. Drugs are typically delivered via a dart fired from a distance. After darting, a period of induction follows, during whichthe drug takes effect. During this time, close monitoring of the animal's vital signs is essential. Once sedated, the animal can be positioned for the procedure and monitored throughout. Reversal drugs are administered at the end to allow for a safe recovery.

8.6.2. Physical Restraint

This method is generally discouraged for Bovidae due to the high risk of injury to both the animal and the handlers. It's stressful for the animal and can lead to complications. Physical restraint may only be used as a last resort in specific situations where darting is impossible and by highly experienced personnel.

Important Considerations

- **Species Variations:** Immobilization protocols will differ depending on the specific Bovidae species. Antelope and gazelle, for instance, may require slightly different drug combinations or handling techniques compared to bison or wildebeest.
- **Expertise is Key:** Immobilization of any wild bovid should only be attempted by a trained veterinarian or wildlife professional with the

experience and knowledge to ensure animal safety and successful intervention.

- **Minimize Risks:** The goal is to achieve rapid and effective immobilization with minimal stress on the animal. This includes having a plan to minimize handling time and ensuring a smooth recovery.
- **Contingency Plans:** Be prepared for complications during immobilization. Having readily available reversal agents and emergency procedures is crucial.
- Minimize the human presence around the animal during recovery to reduce stress.
- Proper handling equipment suited to the specific Bovidae species being immobilized is essential.

Etorphine has been extensively used in captive ruminants (Alford *et al.,* 1974). Xylazine HCl has also been used (Bauditz, 1972).

Etorphine (M-99) dosage (Wallach and Boever, 1983)

Species	Total dose for adult (mg)	Total dose for juvenile (6 Months to 2 Years) (mg)
Greater Kudu	4-7	1-4
Lesser Kudu	3-6	2-3
Sitatunga	3.0-6.0	1.0-3.0
Nyala	1.5-3.0	0.5-1.5
Bush buck	3.0-6.0	2.0-3.0
Eland	5.0-10.0	4.0-6.0
Nilgai	4.0-6.0	1.0-4.0
Buffalo	6.0-10.0	2.0-6.0
Bison	6.0-10.0	2.0-6.0
Wisent	4.0-8.0	2.0-4.0
Yak	3.0-8.0	2.0-4.0
Domestic Cattle	6.0-10.0	2.0-6.0
Hartebeest	2.0-3.0	1.0-2.0
Blesbok	2.0-3.0	1.0-2.0
White-tailed gnu	3.0-5.0	2.0-4.0
Brindled gnu	3.0-5.0	2.0-4.0
Roan antelope	1.5-3.0	1.0-2.0
Sable antelope	3.0-6.0	2.0-4.0
Oryx	5.0-10.0	3.0-7.0
Addax	5.0-8.0	2.5-5.0
Water buck	3.5-5.0	1.5-3.5

Gazelles	2.0-3.0	1.0-2.0
Black buck	2.0-3.0	1.0-2.0
Impala	2.0-5.0	1.5-2.0
Spring buck	1.5-3.0	1.0-2.0
Musk ox	1.5-4.0	1.0-2.0
Ibex	1.0-3.0	0.5-2.0
Tahr	2.0-4.0	1.0-2.0
Aoudads	2.0-4.0	1.0-2.0
Wild sheep	1.0-3.0	1.0-2.0
Red sheep	1.0-3.0	0.5-1.5
Big horn sheep	2.0-4.0	1.0-2.0
Moutlon	1.0-2.0	0.5-1.5

Sarma et al (1997) used 2 ml of Xyalzine (100mg/ml) and 1ml of Ketamine (100mg/ml) for translocation of Blue Bull (Boselaphustragocamelus) of weighing about 200 Kg. lateral recumbency was achieved after 23 minutes of darting. The animal was translocated and Xyazine antagonist Yohimbine HCl 3ml (10mg/ml) was injected intravenously. The reversal action occurs within 2-3 minutes.

Xylazine has been extensively used for the immobilization of captive exotic ruminants and for the capture of free-ranging ruminants. There is no excitement phase with xylazine. Normally, there is an initial drop to the ground. The induction time is 1-15 minutes if given by intramuscular route, and the peak effect occurs at 15-60 minutes. Acomplete recovery takes several hours. Xylazine (0.4 mg/kg) is used in a mixture with M-99 (2.0 μg/kg) or fentanyl citrate (0.5 mg/kg). Both diprenorphine (2 mg) and doxapram(14.0 mg) were used as antagonists.

Xylazine dosages (Bauditz, 1972)

Species	**Xylazine (mg/kg of body weight)**	
	Sedation	**Immobilization**
Domestic ox (*Bos primigenius taurus*)	0.05-0.2	0.3-0.6
Yak (*Bos mutus*)	0.3	0.6-1.0
American Bison (*Bison bison*)	0.1-0.3	0.6-1
European bison (*Bison bonasus*)	0.5	2-3
Chamois (*Rupicapra rupicapra*)		2-3
Ibex (*Capra ibex*)	0.5-1	3-4
Musk ox (*Ovibosmoschatus*)	<0.5	0.5-1.5
Kara Tau sheep (Wild Mountain sheep)	1-1.5	
Water buck (*Kobus ellipsiprymnus*)	1	1

Wild ox	0.1-0.3	0.3-1
Banteng(*Bos javanicus)* orTembadau		0.5-1.5
Gayal		1-1.5
Impala	1	3
Springbok or springbuck (*Antidorcas marsupialis*)	1	1-3
Black buck	1	3
Brindled gnu	1	1.5-3
White tailed gnu	1	1.5-3
Hartebeest (*Alcelaphusbuselaphus*) also known as kongoni or kaama		2
Greater kudu	1	3
Maxwell's duiker	1	2
Sitatunga	1.5	3
Eland	1	3
Nilgai	1	3

8.7. Immobilization of Deer

Deer are common large mammals, sometimes requiring immobilization for various reasons like relocation, disease treatment, or research. Here's a breakdown of the two main methods used for deer immobilization:

8.7.1. Chemical Immobilization

This is the preferred method due to its control and efficiency. Here's a closer look:

- **Drugs:** A veterinarian will choose a specific drug combination based on the deer's size, species, health, and the intended procedure. Factors like age and temperament also influence drug selection. Common drug classes used include opioids, sedatives, and muscle relaxants.
- **Delivery:** Drugs are typically delivered via a dart fired from a distance using a specialized rifle. This minimizes stress on the deer and allows for intervention in less controlled environments like forests.
- **Induction and Monitoring:** After the dart hits, it takes some time for the drug to take effect (the induction phase), lasting around 5-15 minutes. A trained professional will closely monitor the deer's breathing, heart rate, and reflexes during this time. Once sedated, the deer is carefully positioned for the procedure and monitored throughout the immobilization period.

- **Reversal:** After the procedure is complete, an antagonist drug is administered to reverse the effects of the immobilization drugs, allowing the deer to recover safely.

8.7.2. Physical Restraint

Physical restraint is rarely used for deer due to the high risk of injury to both the animal and the handlers. It can only be considered in specific situations and by experienced personnel:

- **Limited Use:** This method should only be attempted as a last resort and only by highly experienced professionals due to the significant risk of stress and injury to the deer.
- **Stressful for Animals:** Physical restraint can be very stressful for deer, increasing the risk of complications like muscle damage or heart failure.
- **Potential Injuries:** The struggle to restrain the deer can lead to injuries for both the animal and the handlers.

8.7.3. Important Considerations for Deer Immobilization

- **Training and Expertise:** Immobilization of any wild deer should only be attempted by a trained veterinarian or wildlife professional with the necessary experience and knowledge of deer biology and safe handling techniques.
- **Minimize Stress and Risks:** The goal is to achieve rapid and effective immobilization with minimal stress on the animal. This includes having a plan to minimize handling time and ensure a safe recovery.
- **Species Variations:** The specific immobilization techniques may vary slightly depending on the deer species (e.g., white-tailed deer, mule deer). A veterinarian will consider these variations when planning the procedure.
- Having a contingency plan in case of complications during immobilization is crucial.
- Minimizing human presence around the deer during the recovery phase can help reduce stress and promote a smooth awakening.
- Proper handling equipment suited to the size and species of the deer being immobilized is essential.

Etorphine HCl (M-99) was successfully used to immobilize and capture both captive and free-ranging white-tailed deer *(Odocoileus virginianus)*. The best response occurs in the 3.0-6.0 mg range. A little higher dose revealed some dangerous effects seen in the animals where antagonists were not used. The drug is administered by automatic projectile syringes (Wolf, 1970).

Etorphine HCl (M-99) was used to immobilize elk or wapiti*(Cervas canadensis),* which is the second largest species within the deer family, Cervidae. An intramuscular injection of 4 mg is given. The sequence of events during immobilization and surgery isdescribed below (Wolf and Swart, 1970).

Time	**Events**
0 second	Intramuscular injection of 4 mg of etorphine HCl (M-99) was given in hip of adult cow elk having estimated weight of 450 lb.
2 min	Ataxic
2 min 50 sec	Recumbent
10 min 10 sec	1 mg of Etorphine HCl (M-99) administered intravenously. Corneal reflex impaired but detectable
16 min 15 sec	1 mg etorphine HCl (M-99) administered intravenously
17 min 30 sec	Moderate muscle tremors, postural and corneal reflex absent
19 min 30 sec	2 ml promazine (50 mg/kg injection, administered intravenously)
21 min 00 sec	2 ml promazine (50 mg/kg injection, administered intravenously)
31 min 30 sec	Initial incision to operation
1 h 3 min 00 sec	Operation completed
1 h 4 min 30 sec	Intravenous injection of 12 mg diprenorphine
1 h 43 min 00 sec	Elk lifted her head and moved legs
1 h 44 min 00 sec	Sternal recumbency and chewing movements.Elk appeared alert.
1 h 44 min 35 sec	Elk stood and appeared fully recovered from the effect of etorphine HCl (M-99)

The optimal dosage for certain captive and free-ranging deer is presented below (Harthoorn, 1966).

Species	**Etorphine mg/4.5 kg**	**Diprenorphine mg/4.5 kg**
Fallow Deer	0.98 (0.23-12.0)	1.96 (0.46-24.0)
Moose *(Alces alces)*	0.98 (0.23-12.0)	1.96 (0.46-24.0)
Tule wapiti	0.98 (0.23-12.0)	1.96 (0.46-24.0)

The xylazine and ketamine combination is an ideal immobilizing agent. Xylazine-ketamine combination on pre-atropinized Mule deer *(Odocoileus hemionus).* Atropine (0.4 mg/kg of body weight) was administered 15 minutes before administering the combination of ketamine (11 mg/kg of body weight) and xylazine (0.22 mg/kg of body weight) administered intramuscularly. Within 10 minutes, complete anesthesia was achieved. A xylazine-etorphine combination can also be used (Jones, 1984).

Chemical immobilization of deer (Jones, 1984)

Species	Suggested drug
Moose or Elk *(Alces alces)*	2.5-4.0 mg etorphine + 4.0-80 mg xylazine
Wapiti or Elk *(Cervus canadensis)*	3.0-4.0 mg etorphine + 15-20 mg acepromazine
Pere David's deer and Sambar	2.5-4.0 mg etorphine + 10-15 mg acepromazine
Swamp deer	2.0-2.5 mg etorphine + 4.5-60 min xylazine
Reindeer	2.0-2.5 mg etorphine + 30-150 mg xylazine
Axis deer	2.0-3.5 mg etorphine + 10-15 mg acepromazine
White tailed deer	1.5-2.0 mg etorphine + 20-40 mg xylazine
Timor deer	1.5-2.0 etorphine + 7-10 mg acepromazine
Hog deer	1.0-1.5 mg etorphine + 5-7 mg acepromazine
Pudu deer	0.4-0.6 mg etorphine + 3-5 mg xylazine
Reeve's muntjac or Chinese muntjac	0.8-1.1 mg etorphine + 4-05 mg acepromazine
Chinese water deer	0.9-1.2 mg etorphine + 4-5 mg acepromazine
Roe deer	1.0-1.2 mg etorphine + 15-20 mg xylazine
Fallow deer	1.2-2.0 mg etorphine + 35-70 mg xylazine
Sika deer	2.0-3.5 mg etorphine + 9-15 mg acepromazine
Red deer	2.0-4.5 mg etorphine + 9-20 mg acepromazine

Xylazine should be used cautiously in an inbred population. The experiments were conducted on an inbred red deer population. The 3-4 mg/kg manufacturers recommended dosage produced profound coma. The correct dosage was found to be 0.1-0.2 mg/kg (Fletcher, 1974).

Xylazine dosage for deer (Bauditz, 1972)

Species	Sedation (mg/kg)	Immobilization (mg/kg)
Fallow deer *(Dama dama)*	1-2	5-8
Indian Hog deer *(Axis porcinus)*	1-2	3-4
Axis deer *(Axis axis)*	1-2	3-4
Sika deer *(Cervus nippon)*	2	3-4
Red deer *(Cervus elaphus)*	1-2	3-4
Roe deer *(Capreolus capreolus)*	0.5-1	1.5-3
White tailed deer *(Odocoileus virginianus)*	0.5-1	3-4
Elk *(Alces alces)*	0.5	1.5
Reindeer or Caribou *(Rangifer tarandus)*	0.5	2

Carfentanil and xylazine combinations are used in immobilizing moose *(Alces alces)* (Seal *et al.,* 1985). The 3-4 mg of carfentaniland 100-175 mg of xylazine were darted using 3 cc dart syringes. Immobilization time ranges from 2.5 to 6 minutes. Reversal was done after 30-90 minutes of darting using naloxone and diprenorphine given intramuscularly and intravenously. The recovery time varied from 10 minutes to 3 hours. Quite undisturbed moose can be immobilized with 3 mg of carfentaniland 100 mg or less of xylazine. Carfentanil and xylazine were administered simultaneously using commercial syringe darts fired from a helicopter to desert big horn sheep, tole, elk, and a wild horse (Jessup *et al.,* 1985).

For Bighorn sheep,the dosage is: carfentanil- at 0.044±0.02 mg/kg and xylazine at 0.190±0.04 mg/kg of body weight. The induction time is 6.3±3.2 minutes. The antagonist is diprenorphine, administeredintravenouslyat 0.299 mg/kg of body weight. Nalaxone dose is 7.68 mg/kg body weight administered intravenously. Yohimbine is administered intravenously at0.165 mg/kg of body weight to antagonize the action of xylazine.

For Elk, the dosage is: Carfentanil- at0.019±0.09 mg/kg and xylazine at 0.235±0.08 mg/kg of body weight. The induction time was 9.2±6.8 minutes. The antagonists arediprenorphine at 0.105 mg/kg body weight, Nalaxoneat 1.03 mg/kg body weight. Yohimbine+Diprenorphine+Nalaxonemay be administered at 0.5 mg/kg+at0.065 mg/kg+at0.48 mg/kg body weight, administered intravenously, to reverse the effects of xylazine and carfentanil. The reversal time is 7.5 minutes. There is slow absorption of carfentanilfrom fat depots and slower metabolic breakdown. Nalaxone has a relatively shorter half-life. So low doses of Nalaxone induce a re-narcotization by carfentanil

Antagonists to xylazine and ketamine induced anaesthesia

Combining ketamine and xylazine at ratios of10-15 parts to1 part, respectively, is sufficient in deer with adequate muscle relaxation and acceptable cardiac and respiratory rates. Smooth and rapid induction occurs (Jessup, 1983). The ketamine dosage is 5.8-14.5 mg/kg (9.2 mg/kg), and thexylazine dosage is 0.44-0.92 mg/kg (0.73 mg/kg). Yohimbine is administered intravenously at 0.125/kg of body weight withmarked respiratory stimulation. The mean recovery time is 8.2 minutes after administration of the injection.

Yohimbine, an α_2-adrenergic blocking agent, appears to be capable of reversing the dissociative effect of ketamine and the sedative effect of xylazine, which induced prolonged sedation in white-tailed deer (*Odocoileus virginianus*) and was reversed by yohimbine (Hsu, 1984).

In mule deer,xylazine was administered at 3.7±1.2 mg/kg of body weight. The induction time was 13 minutes and the recovery time was 268±76 minutes. The action of xylazine was antagonized by yohimbine. The time of standing after giving yohimbine was 4±5 minutes, and it reversed the bradycardia and respiratory depression induced by Xylazine (Jessup *et al.,* 1985).

Yohimbine and 4-aminopyridine were used to antagonize the effect of xylazine. In North American Cervidae,yohimbine was administered at 0.15 mg/kg and 4-aminopyridine at 0.26-0.29 mg/kg body weight, administered intravenously. It produced arousal within 1-15 minutes (Renecker and Olsen, 1985).

Yohimbine is used to antagonize ketamine-xylazine-produced-anesthesia in white-tailed deer. The deer recovered within 2.0 minutes, stood up in 6.0 minutes, and walked away after 9.5 minutes after the injection of yohimbine (Mech *et al.,* 1985)

Yohimbine and 4-aminopyridine wereused to antagonize the effect of xylazine HCl in captive Wapiti. The animal attained the ability to respond to auditory stimuli in an average time of 4±2.1 minutes. Total recovery shortened by an average of 2-5 hours (Renecker and Olsen, 1985).

Bison and Wildebeest: Research papers often explore immobilization methods for these larger Bovidae members. Look for studies on "American bison immobilization" or "Wildebeest immobilization techniques".

Cattle, Sheep, and Goats: These domesticated Bovidae may have established veterinary procedures for immobilization during medical procedures. Consult a large animal veterinarian for details.

9

Chemical Restraint of Equids, Tapirids, Rhinoceroses and Elephants

9.1. Classification: Order Perissodactyla

The order Perissodactyla, also known as odd-toed ungulates, is a group of herbivorous mammals characterized by having one or three toes on each foot. They are distinguished from even-toed ungulates (Artiodactyla) by this anatomical feature. Even-toed ungulates, such as pigs, cows, and deer, have an even number of toes (two or four) on each foot. Some of the well-known members of this order include horses, zebras, asses, tapirs, and rhinoceroses.

The name "Perissodactyla" comes from the Greek words "perissos," meaning "odd," and "daktylos," meaning "finger or toe," referring to their unique foot structure. Unlike even-toed ungulates (Artiodactyla), which bear weight on their third and fourth toes, perissodactyls primarily bear weight on their third toe.

There are around 17 living species of perissodactyls, classified into three families:

9.1.1. Family Equidae

This family includes the familiar horses, donkeys, and zebras. They are known for their long legs, powerful hooves, and grazing habits. The family Equidae includes Mangolian wild horses (*Equus przewalski*), Grevy's Zebra or Imperial Zebra (*Equus grevyi*), Common Zebra (*Equus burchelli*), Mountain Zebra (*Equus zebra*), and Plains Zebra (*Equus quagga*).

Horse Donkey

9.1.2. Family Rhinocerotidae

Rhinoceroses are the largest living perissodactyls and are easily recognizable by their horns and thick skin. The family Rhinocerotidae includes white rhinoceros or square-lipped rhinoceros (*Ceratotherium simum*), black rhinoceros (*Diceros bicornis*), Indian rhinoceros or greater one-horned rhinoceros (*Rhinoceros unicornis*), Javan rhinoceros (*Rhinoceros sondaicus*), and Sumatran rhinoceros or Asian two-horned rhinoceros (*Dicerorhinus sumatrensis*).

Rhinoceros

9.1.3. Family Tapiridae

Tapirs are pig-like mammals found in Central and South America and Southeast Asia. They have a short, prehensile snout and four toes on their front feet. The family Tapiridae includes Brazilian Tapir (*Tapirusterrestris*), Little Black Tapir (*Tapirus kabomani*), American Tapir (*Tapirus terrestris*), and Mountain or Wolly Tapir (*Tapirus pinchaque*).

Tapir

Perissodactyls have a long and rich evolutionary history, dating back to the Eocene epoch (56 million to 33.9 million years ago). They were once a much more diverse group than they are today, and they included many large and bizarre-looking creatures.

Some of the interesting characteristics of perissodactyls include:

- Their feet are adapted for running. The weight of the animal is borne on the middle toe, which is the strongest and most hoof-like. The other toes are either reduced or absent.
- They have a long digestive system, which is necessary for breaking down the tough plant material that they eat.
- They have a good sense of smell, which they use to find food and avoid predators.
- Many perissodactyls are social animals and live in herds or groups.

Today, perissodactyls are an important part of many ecosystems around the world. Unfortunately, several species of perissodactyls are threatened with extinction due to habitat loss, hunting, and poaching.

9.2. Anaesthesia for the Equidae

Anaesthesia for the Equidae family, which includes horses, donkeys, and zebras, is a critical component of many veterinary procedures. It allows veterinarians to perform surgeries, examinations, and other procedures safely and humanely.

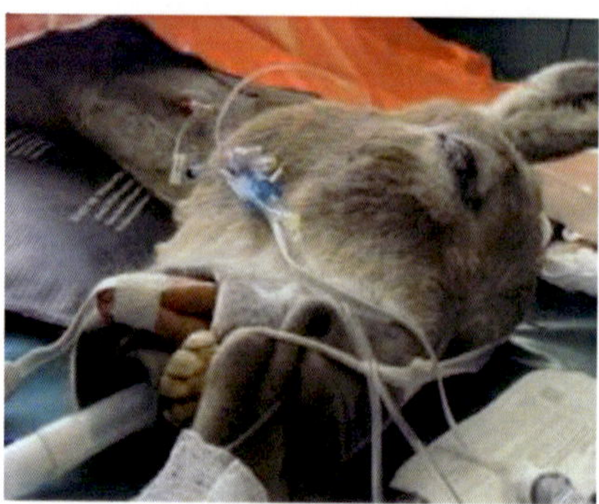

Donkey anaesthesia

There are two main types of anaesthesia used in equine medicine:

- **General anaesthesia:** This renders the horse completely unconscious and unable to feel pain. It is typically used for major surgeries or procedures that require complete immobilization.
- **Local anaesthesia:** This numbs a specific area of the body, allowing the horse to remain conscious during the procedure. It is typically used for minor surgeries or procedures in a small area.

The specific anaesthetic protocol used for an equine patient will depend on a number of factors, including:

- The type of procedure being performed
- The age and health of the horse
- The horse's temperament

Here's a general overview of the process of anaesthesia for an Equidae:

- **Pre-anaesthetic assessment:** Before any anaesthesia is administered, the veterinarian will perform a thorough pre-anaesthetic assessment. This will include a physical examination, a review of the horse's medical history, and blood tests.
- **Premedication:** The veterinarian may administer premedication to help relax the horse and reduce the risk of complications during anaesthesia.
- **Induction of anaesthesia:** General anaesthesia is typically induced by administering an intravenous injection of a fast-acting anaesthetic agent.
- **Maintenance of anaesthesia:** Once the horse is unconscious, the veterinarian will maintain anaesthesia with a combination of inhalational and intravenous anaesthetic agents.
- **Monitoring:** Throughout the procedure, the veterinarian will closely monitor the horse's vital signs, such as heart rate, respiratory rate, and blood oxygen levels.

- **Recovery:** After the procedure is complete, the horse will be allowed to recover from anaesthesia in a quiet, padded stall.

The most common anaesthetic agents used in equine medicine are:

- **Guaifenesin:** A muscle relaxant that is often used in combination with other anaesthetic agents.
- **Ketamine:** A dissociative anaesthetic that produces a trance-like state.
- **Propofol:** A short-acting anaesthetic agent that is often used for the induction of anaesthesia.
- **Sevoflurane:** An inhalational anaesthetic agent that is commonly used for the maintenance of anaesthesia.

Anaesthesia is a safe and effective procedure when performed by a qualified veterinarian. However, there are always some risks associated with anaesthesia, such as respiratory depression, allergic reactions, and drug overdose. These risks can be minimized by careful pre-anaesthetic assessment, proper monitoring, and the use of appropriate anaesthetic agents and techniques.

An etorphine and acepromazine combination has been used for prolonged anesthesia in a zebra *(Equus burchelli)*. The dose was 5 mg and 15 mg, respectively, given intramuscularly. In turn, a 0.5 mg dose of etorphine was given, i.e., after maintaining anesthesia. After 30-45 minutes, the animal was recovered. Acepromazine increases the heart rate due to the potentiation of β-adrenergic effects and α-adrenergic blockade. This decreases the etorphine-induced hypertension (David *et al.*, 1984).

Bogan *et al.* (1978) recommended the use of acepromazine at a dose rate of 0.1 mg/kg of body weight given intramuscularly 20 minutes before or intravenous administration 10 minutes before etorphine. This caused a reduction in hypertension and the easy administration of etorphine.

In the horse, normal casting doses of the etorphine-acepromazine mixture produced a marked tachycardia and rise in blood pressure with the increase in heart rate. This can be reduced by premedication with propranolol (Daniel *et al.*, 1972).

The fall in pH (arterial blood) results after a short chase of only a few km proved lethal in untreated zebra *(Equus burchelli)* during immobilization (Harthoorn, 1974). Acidemia was rectified by an infusion of alkaline fluid, and these treated animals survived. For Equidae, etorphine-acepromazine (0.25 ml/50 kg) was administered intramuscularly. It is antagonized by an equal dose rate of diprenorphine.

9.3. Anaesthesia for Rhinocerotidae

Anesthetizing a rhinoceros is a complex and challenging procedure due to their immense size, physiology, and sensitivity to certain medications. Here's a breakdown of anaesthesia for these massive mammals:

9.3.1. Challenges of Rhino Anaesthesia

- **Large size:** The sheer size of rhinos (up to 3,500 kg for some species) makes delivering the right amount of anaesthesia difficult.
- **Respiratory issues:** Rhinos are prone to respiratory depression under anaesthesia, requiring close monitoring and potential ventilator support.
- **Opioid sensitivity:** They are particularly sensitive to opioids, necessitating careful selection and dosage of these drugs.

9.3.2. Common Anaesthetic Regimens

- **Combination approach:** Typically, a combination of drugs is used for rhino anaesthesia. This might involve:
  - **Opioids:** Powerful opioids like etorphine hydrochloride are often used for initial sedation, but due to the sensitivity mentioned earlier, they need to be combined with other drugs.
  - **Alpha-2 adrenoreceptor agonists:** Drugs like detomidine or azaperone help with sedation and pain relief.
  - **Other medications:** Depending on the procedure, additional medications like ketamine (anaesthetic), butorphanol (pain reliever), or guaifenesin (muscle relaxant) might be used.
- **Inhalational Anaesthetics:** For extended procedures, inhalational anaesthetics like isoflurane or sevoflurane might be used, but maintaining the equipment for these in field settings can be difficult.

9.3.3. Delivery Methods

- **Darting:** In most cases, the initial anaesthetic drugs are delivered via a dart fired from a distance. This minimizes stress on the animal and allows for safer administration.

Importance of Experienced Team

Due to the complexities involved, rhino anaesthesia should only be performed by a highly trained and experienced veterinary team. They will have the expertise to choose the right drugs, deliver them safely, and closely monitor the animal throughout the procedure.

Additional Considerations

- **Pre-anaesthetic assessment:** A thorough health check is crucial before anaesthesia to identify any underlying conditions that could complicate the process.
- **Monitoring:** Vital signs like heart rate, respiration, and oxygen levels are continuously monitored during anaesthesia to ensure the animal's safety.
- **Recovery:** Rhinos need a quiet and controlled environment to recover from anaesthesia, with close supervision until they are fully awake and stable.

Overall, anesthetizing a rhinoceros is a specialized procedure requiring a combination of powerful drugs, skilled professionals, and careful monitoring to ensure the animal's well-being. Etorphine-acepromazine (0.15 mg/100 kg of body weight: 0.01 mg/kg) administered intramuscularly is very effective. Etorphine-ketamine (1.0 mg/500 kg of body weight: 0.5 mg/kg) is administered intramuscularly. Best reversed with nalorphine.

Sarma *et al.* (1997) used xyalzine to restrain and treat an injured rhinoceroses in Kaziranga National Park. A total of 1500 mg of xyalzine HCl (100 mg/ml) was administered using a projectile syringe in a Disinject system in two shots. 10 ml in the brachiocephalicus muscle over the neck and 5 ml in the quadriceps femoris muscle in the hind quarters. The animal attained lateral recumbency within 15 minutes.

9.4. Anaesthesia for Hippopotamidae

The family Hippopotamidae, also known as hippopotamids, consists of two extant species of large, semiaquatic mammals:

- The common river hippopotamus (*Hippopotamus amphibius*) is the larger of the two species and the second-largest land mammal after the elephant. They are found in sub-Saharan Africa and are recognizable by their massive size, barrel-shaped bodies, and large heads.

Common hippopotamus

- The pygmy hippopotamus (*Choeropsis liberiensis*) is much smaller than the common hippopotamus and is about the size of a large boar. They are found in forested areas of West Africa, and they are much more reclusive than common hippos.

Pygmy hippopotamus

Despite their resemblance to pigs, their closest living relatives are actually cetaceans (whales, dolphins, and porpoises).

Anesthesia for hippopotamuses (Family Hippopotamidae) is a complex and dangerous procedure. Here's a breakdown of the challenges and considerations:

9.4.1. Challenges of Anaesthesia in Hippos

- ***Large size:*** Hippos are the second-largest land mammals after elephants, making anesthesia administration and monitoring difficult.
- ***Semi-aquatic nature:*** Hippos spend a significant amount of time underwater, posing a drowning risk during anesthesia.
- ***Aggressive temperament:*** Hippos are highly territorial and aggressive, making handling them during anesthesia recovery risky.
- ***Physiology:*** Hippos have unique physiological adaptations for their aquatic lifestyle, including thick skin, dense muscle mass, and a slowed metabolism, all of which can complicate anesthesia.

9.4.2. Considerations for Anaesthesia in Hippos

- ***Pre-anesthetic assessment:*** A thorough health evaluation is crucial to identify any underlying conditions that could be exacerbated by anesthesia.
- ***Immobilization techniques:*** Immobilization methods like darting or crate capture are used to minimize stress and injury during the procedure.
- ***Anaesthetic drugs:*** A combination of drugs is typically used to achieve adequate anesthesia, analgesia (pain relief), and muscle relaxation.

- ***Monitoring:*** Close monitoring of vital signs like heart rate, respiration, and blood oxygen levels is essential throughout the procedure.
- ***Recovery:*** Hippos require specialized care during recovery to ensure they breathe properly and maintain body temperature.

Overall, anesthesia in hippos is a high-risk procedure that should only be performed by experienced veterinary teams with specialized equipment and expertise.

Here are some additional points to consider:

- Anesthesia in hippos is typically only used for essential medical procedures, such as dental work, abscess treatment, or wound repair.
- Minimally invasive techniques are preferred whenever possible to reduce the need for anesthesia.
- Research is ongoing to develop safer and more effective anesthesia protocols for hippos.

9.4.3. Challenges of Anaesthesia in Hippos

- Owing to their aggressive temperament, they have difficulties preventing dehydration and hyperthermia if immobilized on land. There are difficulties in preventing drowning if these animals are administered drugs while swimming. The hippopotamidae family presents severe restraint problems.
- Phencyclidine-acepromazine (0.5 mg/kg-0.2 mg/kg, administered intramuscularly)
- Xylazine alone, at1.3 mg/kg is considered to be the safest agent if the animals are in or near water. Etorphine (4-5 mg/300 kg) has been recommended for restraint on land (Green, 1982).

9.4.4. Reasons for Anaesthesia in Hippos

There are several reasons why a hippo might require anesthesia:

- **Dental care:** Hippos' teeth continuously grow throughout their lives. Sometimes, these teeth can become overgrown or misaligned, causing discomfort or interfering with eating.

Hippopotamus teeth

- **Wound treatment:** Hippos can sustain injuries from fights with other hippos or encounters with predators. Anaesthesia is necessary for cleaning and treating these wounds.
- **Translocation:** In rare cases, hippos may need to be relocated due to habitat loss or human-wildlife conflict. Anaesthesia can help reduce stress during transport.

9.5. Anesthesia for Trapiridae

Tapirs belong to the family Tapiridae, within the order Perissodactyla, and are distinguished from the Artiodactyla by their foot morphology and digestive system. Despite their large size, tapirs have had a reputation for being docile in captive environments and extremely elusive as free-ranging animals. General anesthesia is maintained by inhalation anesthetics or total intravenous anesthesia (TIVA). Once anesthetized, tapirs are maintained in sternal or lateral recumbency. Anesthetizing tapirs, members of the Tapiridae family, also requires a specific approach due to their unique physiology and temperament. Here's what you need to know about anaesthesia for tapirs:

9.5.1. Anaesthetic Considerations for Tapirs

- Similar to Rhinos: Anaesthesia for tapirs shares some challenges with rhinos, such as respiratory depression. Tapirs are susceptible to breathing difficulties under anaesthesia, requiring close monitoring and potential oxygen supplementation.
- Physiological Differences: Unlike rhinos, tapirs are not particularly sensitive to opioids. However, their smaller size compared to rhinos necessitates careful medication dosage.
- Stressful procedures: Tapirs can be easily stressed by capture and handling, making pre-medication and a calm environment crucial for a smooth procedure.

9.5.2. Common Anaesthetic Regimens for Tapirs

- Combination approach: Similar to rhinos, a combination of drugs is typically used:
  - *Opioids:* Butorphanol is a common choice due to its lower sensitivity compared to rhinos.
  - *Alpha-2 adrenoreceptor agonists:* Drugs like detomidine or xylazine provide sedation and pain relief.
  - *Other medications,* such as ketamine (a dissociative anaesthetic), or midazolam (a sedative) might be used depending on the procedure.
- Inhalational Anaesthetics: Isoflurane or sevoflurane may be used for extended procedures, but maintaining the equipment can be challenging in field settings.
- Delivery Methods:
- *Remote injection:* Drugs are often delivered via remote injection darts to minimize stress on the animal.
- *Intravenous injection:* In some cases, intravenous injection through the jugular vein may be used for more precise medication administration.
- Importance of an Experienced Veterinarian: As with rhinos, a qualified veterinarian experienced in tapir anaesthesia is essential. They can choose the most suitable drugs, administer them safely, and monitor the animal's vital signs throughout the procedure.

9.5.3. Additional Considerations

- Pre-anaesthetic assessment: A thorough health check is necessary to identify any underlying conditions that could pose risks during anaesthesia.
- Stress management: Minimizing stress through a calm environment and handling techniques is crucial for a safe procedure.
- Recovery: Close monitoring is required during recovery until the tapir is fully awake and stable.
- Remember: Anaesthesia for tapirs requires a tailored approach considering their physiology and potential for stress. Consulting a veterinarian experienced in tapir anesthesia is vital for ensuring the animal's safety and well-being during any procedure.

9.5.4. Restraint and Handling

Tapirs have traditionally been managed through "direct contact"; however, reports of severe or lethal injuries to caretakers highlight the risks of manual restraint. The current Association of Zoos and Aquariums (AZA) standards recommend that tapirs be managed with "protected contact." The only acceptable type of physical immobilization, which requires training, is the use of a large animal chute. Depending on the disposition of individual animals, some tapirs may be trained with positive reinforcement to be moved, to be prompted to present body parts, and to stand still for biologic sample collection or intravenous catheterization. Tapirs may also be "scratched-down" to lateral recumbency, with a coarse horse brush or outdoor broom used to stroke the animal's dorsum, neck, and the lateral and abdominal walls. Although some have used "scratch-downs" for physical examination, administration of injections, and even repeated blood collection, the level of immobilization induced by this "state" should not be exaggerated (Viviana Quse and Renata Carolina Fernandes-Santos, 2014).

The success of the chemical restraint of wild and captive tapirs depends on careful planning. During the planning stages, it is important to consider the following:

1. Characteristics of the anatomy, metabolism, and physiology of the species.
2. The environmental conditions of the location where the capture will take place.
3. The capture method that will be employed.
4. The available equipment that might be used during the capture process.
5. Detailed knowledge of the pharmacology, adverse effects, and counter-indications of the drugs that will be used for the chemical restraint.
6. Estimates of the time required to carry out all the procedures during the manipulation of the animal, including but not limited to: radio-tagging, installation of microchips, collection of biological samples, and clinical evaluation.
7. The possibility of unexpected events interrupting or interfering with the chemical restraint.
8. The need to check different physiological parameters of the tapirs during their recovery from anesthesia.

Anaesthetic protocols to be used to immobilize tapirs in the field must satisfy the following criteria:

1. They must provide a rapid induction to avoid predation, trauma, or accidental drowning.
2. They should be fully reversible to ensure a rapid recovery.

3. They should have a relatively wide margin of safety since pre-capture weights are difficult to obtain in the field.
4. They should provide adequate immobilization to ensure sedation, as long as it is needed to accomplish the desired procedures.
5. They should be relatively inexpensive.

The recent increase in use of α_2-adrenergic agonists, alone or in combination with other drugs, has decreased the need for ultrapotent narcotics such as etorphine and carfentanil. The preferred anaesthetic protocol depends on both the environmental setting (i.e., captive or free-ranging) and the individual's disposition. It is important to establish a quiet environment while working with these animals. If stress is unavoidable (i.e., transport, trauma, or disease), preanesthetic medication is recommended. The success of some anaesthetic protocols used for free-ranging tapirs depends on the animal's degree of relaxation. Tapirs often retreat into water when threatened, so this should be taken into consideration during immobilizations. It is recommended that the animal be fasted for 24 hours to minimize the potential for regurgitation and minimize the pressure on the diaphragm from the GI tract. Protocols should emphasize the maintenance of euthermia through the heating of enclosures in cold climates or cooling the animal in hot weather. Because tapirs are tropical animals that are rarely exposed to full sunlight, they should not be immobilized in full sun.

Summary of protocols used to immobilize free ranging tapirs (Viviana Quse and Renata Carolina Fernandes-Santos, 2014)

Species	Animal Status	Anesthetics	Comments
Tapiruspinchaque	Captive	Carfentanil (5.4 microgram per kilogram [µg/kg]), ketamine (0.26 milligram per kilogram [mg/kg]) and xylazine (0.13 mg/kg,) intramuscularly [IM])Reversal: yohimbine (0.2 mg/kg, intravenously [IV]); and naltrexone (100–200 mg/kg, half IV, half subcutaneously [SC])	Six immobilizations of tapirs (1 female, 3 males; 1 juvenile male immobilized 3 times) for footwork, gastrointestinal endoscopy, reproductive surgery[52]
Tapirus indicus	Captive	Butorphanol (0.15 mg/kg) and detomidine (0.05 mg/kg) OR xylazine (0.3 mg/kg, all IM); use ketamine if needed (0.5 mg/kg, IV) Reversal: naloxone and yohimbine (0.2–0.3 mg/kg, IV)	Nineteen immobilizations of Malayan and mountain tapirs.
Tapirus bairdii	Captive	Butorphanol (0.15–0.2 mg/kg) and detomidine (0.06 mg/kg); ketamine (1–2 mg/kg) was used to reach or maintain recumbency and light anesthesiaBoluses of propofol (0.2–2.0 mg/kg per bolus) were used to effect as needed	Report of 11 males anesthetized for semen collection by electroejaculation in three captive institutions in Panama,[16] but authors (personal communication) report a similar usage in similar number of females has the same effects, and that anesthesia was antagonized with naltrexone and yohimbine.
Tapirus indicus	Captive	Butorphanol (80 mg, IM) andXylazine (120 mg total, IM) OR detomidine (12 mg total, IM)Reversal: naltrexone (200 mg total, IM), tolazoline (1400 mg total, IM)	Estimated body weight: 340 kg; repeated immobilizations with light anesthesia for the diagnosis and treatment of oral squamous cell carcinoma.

Tapirus terrestris	Captive	Detomidine (0.03mg/kg, orally [PO]), 20 minutes later, carfentanil (1.85 μg/kg, PO)	One animal was repeatedly immobilized for wound management; a variety of combinations of α_2-agonist, etorphine, or carfentanil were used, but eight immobilizations with detomidine/carfentanil, PO, were most useful.
Tapirusbairdii	Attracted wild tapirs to bait stations	Total dosage for a 200–300 kg animal: 40–50 mg of butorphanol and 100 mg of xylazine in the same dartAdditional ketamine (187 ± 40.86 mg/animal) or constant rate infusion of propofol (50–200 μ/kg/min), administered IV. Reversal: Naltrexone 50 mg, with 1200 mg of tolazoline in the same syringe IM; no sooner than 30 minutes from last administration of ketamine.	Administered to animals from a tree blind via a dartThe animals had been habituated to come to bait for several days and thus were relatively calm when darted.
Tapirusterrestris	Tapirs were captive, semi-captive or wild	Ketamine (3.5–4 mg/kg) and xylazine (2–2.2 mg/kg) IM, supplemented with ketamine (1.4 mg/kg) IMReversal: Tolazoline (4 mg/kg)	Administered using darts projected by a blowpipe, or IV using syringe.
Tapirusterrestris	Wild tapirs captured in pens or pit-falls	Butorphanol tartrate (0.15 mg/kg) with medetomidine (0.03 mg/kg) IM, in same dartReversal: Atipamezole (0.06 mg/kg) with naltrexone (0.6 mg/kg) in same syringe, IV	Adequate immobilization for radio collaring, and biologic sampling.

Tapirus terrestris and Tapirus pinchaque	Wild tapirs captured in pens or pit-falls, or immobilized by dart	Dosages were calculated using allometric scaling: ketamine (0.62–0.41 mg/kg) and atropine (0.025–0.04 mg/kg), and tiletamine-zolazepam (1.25–083 mg/kg), and romifidine (0.05–0.03 mg/kg) or detomidine (0.06–0.04 mg/kg) or medetomidine (0.006–0.004 mg/kg) in the same dart. Reversal: atipamezole (0.06 mg/kg)	Medetomidine produced best results obtaining good muscular relaxation and more stable cardiopulmonary parameters[43]

Tapir displaying a head down, wide stance, typical of the initial effects caused by the induction anaesthetics.

Once the animal fell into sternal recumbency, the veterinarian would climb down from the platform and approach the animal to apply a blindfold, insert gauze in the ears, and begin anesthesia monitoring. While immobilized, the tapirs were fitted with radio collars. In addition, morphometric measurements, and biological samples (blood, feces, hair, skin biopsies, and urogenital swabs) were collected. Biological parameters (heart rate, respiratory rate, direct and indirect blood pressure, body temperature, blood gases, and ECG) were monitored and recorded (Foerster *et al.* 2000).

Some of the aforementioned procedures required that the tapir be manipulated, requiring further sedation to avoid premature arousal. Ketamine (i.e., Ketaset) was utilized in 19 of the 33 immobilizations as a means to either achieve deeper sedation or extend the anaesthetic period. Of the animals that received ketamine, 13 received it intravenously, while 6 received it as an intramuscular injection. An intravenous catheter was placed in the ear vein to administer ketamine.Ketamine was administered in increments of 25 mg (range 25 - 200 mg) and the mean total dose used was 123.50 mg per animal. Propofol (i.e., Rapinovet) was used to provide further sedation in 5 animals. This was administered as a constant-rate infusion at a range of 50 - 200 micrograms/kg/minute based on the estimated weights of each individual. Table 1 illustrates the different doses of propofol used on five tapirs.

Table 1. Different Doses of Propofol Used on Five Tapirs.

Tapir	Estimated Body wt (kg)	Dose of Propofol (mg/kg/mt)	Total Propofol Administered (mg)
Rio	300	0.33	1.8
Leftie	300	0.10	5.7
Roberta	300	0.10	6.8
Flash	300	0.23	10.4
Big Mama	300	0.20	5.4

Naltrexone (i.e., Trexonil) was injected intramuscularly in all immobilized tapirs to reverse the butorphanol. The mean dose of naltrexone utilized was 153.50 mg/animal (range 50 - 300 mg). Either yohimbine (i.e., Yobine) or tolazoline (i.e., Tolazine) was administered intramuscularly to reverse the alpha-2 agonist effects of xylazine. Yohimbine was used at a mean dose of 34 mg/animal, while tolazoline was used at a mean dose of 961.33 mg/animal (range of 400 - 1200 mg). The reversal agents were administered no sooner than 15 minutes after the last administration of ketamine. After the reversal drugs were administered, the cotton gauze was removed from the ears, but the blindfold was left in place. The veterinarian recorded all time-related events, such as time of dart impact, time of first effects, time of sternal recumbency, time of reversal administration, time to return to sternal position, time to return to standing, and total time immobilized.

The first effects of sedation were observed in an average of 4 minutes from dart impact. The first sign associated with sedation was an inability to prehend bananas with their proboscis. Other signs observed consistent with the first effects were ataxia, wide base stance, a head-down posture, penile prolapse in males, and hypersalivation. Animals achieve sternal recumbency in an average of approximately 11 minutes. Animals went down by first lowering their hindquarters and then extending their forelimbs in front of them. The head was the last to be lowered to the ground. During two different immobilizations of the same tapir, the animal assumed a wide base stance but did not lie down until manually pushed over. The animals were anesthetized an average of 45 minutes. Three animals experienced premature arousal. This was defined as an animal assuming a sternal position or standing prior to the administration of reversal agents. Two of these animals did not receive ketamine. The remaining animal had received ketamine intramuscularly.

The average time from reversal administration to return to sternal recumbency was approximately 4 minutes. The average time it took for tapirs to stand was approximately 12 minutes after reversal administration. Animals that received propofol took an average of approximately 8 minutes to return to sternal recumbency after reversal administration. It took them, on average, approximately 12 minutes to return to standing. Often, tapirs returned to eating bananas soon after recovery. The results of the physiologic parameters such as heart rate, respiratory rate, body temperature, SP02, blood gases, and ECG (Paras-Garcia *et al.*, 1996).

9.5.5. Immobilization Protocol

Several anesthetic protocols have been previously developed and used for captive (Janssen *et al.* 1999; Nunes *et al.* 2001; Janssen 2003) and wild tapirs (Hernandez *et al.* 2000; 2001; Mangini *et al.* 2001; Mangini 2007). The specific protocol used will vary depending on a variety of factors, including: capture method, goals of immobilization, desired or feasible drug volume, desired time for induction and recovery, required level of immobilization and muscle relaxation, need for reversibility, project budget, and requirements for safety of the animal and the members of the field team. Several tapir field projects have used anesthetic protocols that have been exhaustively tested in wild tapirs in different areas (Parás-García *et al.* 1996, Hernández-Divers *et al.* 1998; Hernández-Divers *et al.* 2000; Hernández-Divers & Foerster 2001; Mangini *et al.* 2001; Mangini 2007; Medici 2010; Pérez 2013). Researchers planning on using these protocols are strongly advised to consult with an experienced wildlife veterinarian prior to implementing the protocol in the field. In addition, it is highly recommended that the veterinarians who developed and/or have experience with each of these protocols be contacted for further consultation. The conditions under which these protocols are successful should be carefully explored and considered when attempting to apply them to different situations. Further information about the drugs described in this chapter and their effects on animal physiology is available in Appendix 1.

Several protocols have been utilized to immobilize captive tapirs (Janssen *et al.* 1999; Kuehn G. Tapiridae 1986; Paras-Garcia *et al.* 1996). Immobilization protocols that have been reported for free-ranging tapirs include etorphine/acepromazine combinations (Paras-Garcia*et al.*, 1996), butorphanol/xylazine/ketamine combinations (Foerster*et al.*, 2000) and tiletamine-zolazepam and medetomidine combinations (Mangini and Medici, 1998).

Etorphine/acepromazine combinations provide rapid induction but have a low margin of safety and pose a human health hazard. Tiletamine-zolazopam and medetomidine produce prolonged recoveries, which can be dangerous in animals that are not confined.

Butorphanol, xylazine, and ketamine combinations have been utilized to anesthetize horses routinely (Muir and Hubbell, 1991). This protocol provided rapid induction, rapid recovery after reversal administration, adequate immobilization with the addition of ketamine, and a wide safety margin. Tapirs are phylogenetically related to horses and appear to share similar physiology. Furthermore, the protocol described fulfilled the aforementioned requirements.

Typical of alpha-2 agonist and agonist/antagonist narcotic combinations, one problem that could be encountered with this protocol is premature arousal (Muir and Hubbell, 1991). Loud noises or excessive manual manipulation can cause premature arousal. Placing cotton gauze in their ears minimized noise detected by the animals. The addition of ketamine or propofol to the protocol eliminated this problem by allowing further sedation to manipulate the animals and extend the anesthetic period. In a previous study that looked at the anesthetic effects of this protocol, the most consistent anesthetic effects without premature arousal were obtained with 50 mg of butorphanol, 100 mg of xylazine, and 100 - 250 mg of ketamine/animal (Foerster *et al.*, 2000). The lower ranges of ketamine were utilized for procedures that required minor manipulation (rolling to lateral recumbency), whereas the higher ranges were utilized while pushing or pulling the animal. The main disadvantage encountered with the administration of propofol was a prolonged reversal administration-sternal recumbency period when compared with animals in which ketamine alone was used.

This could be due to the different metabolic pathways utilized to break down these two agents. Subjectively, we observed that animals that received propofol took longer to recover and were more likely to be ataxic upon standing than those that did not. The administration of propofol to extend the anesthetic protocol is not recommended unless the tapir can be confined after recovery.

This capture and immobilization protocol proved to be useful to radio collar and collect biological samples from free-ranging Baird's tapirs in Costa Rica. We believe other tapir researchers in other countries can successfully implement the methods presented here. Considering the threatened status of the *Tapiridae* family and the increase in studies requiring the capture of tapirs, it is vital that extreme care be taken to ensure the safety of the animals during these procedures. Table 2 summarizes the recommended anesthetic protocol for the immobilization of Baird's tapirs. These dosages are based on the authors' experience with the protocol discussed throughout the years (Matola *et al.*, 1997).

Table 2. Recommended Anesthetic Protocol for Immobilization of Baird's Tapirs.

Estimated Body wt (kg)	Induction Anesthetic	Supplemental Drugs	Reversal Drugs
300 kg	50 mg Butorphanol	25 - 150 mg Ketamine	50 mg Naltrexone
	100 mg Xylazine		1200 mg Tolazoline

9.6. Classification: Order Proboscidea

The order Proboscidea is a taxonomic group of afrotherian mammals containing one living family, Elephantidae (elephants and mammoths), and several extinct families. Elephants are the largest living land animals on Earth and are easily recognized by their trunks, tusks, large ears, and thick legs.

9.6.1. Family: Elephantidae

Elephantidae is the only surviving family within the order Proboscidea. These are the giants of the land, with thick legs, large ears, and a distinct trunk – a long and muscular nose that serves numerous purposes. They are also known for their tusks, which are elongated upper incisor teeth. Here are some interesting characteristics of Elephantidae:

- ***Trunk:*** The trunk is a prehensile extension of the upper lip and nose, used for breathing, smelling, grabbing food and water, trumpeting, and manipulating objects. It contains over 40,000 muscles and tendons, making it incredibly strong and dexterous.
- ***Tusks:*** Tusks are elongated incisor teeth that continue to grow throughout an elephant's life. They are used for digging, fighting, feeding, and lifting objects.
- ***Ears:*** The large ears of elephants help them regulate their body temperature by fanning themselves with their ears.
- ***Legs:*** Their thick legs provide support for their massive bodies. Elephants are the only mammals with plantigrade feet, meaning their entire foot rests flat on the ground.
- ***Intelligence:*** Elephants are highly intelligent animals with complex social structures. They are capable of self-awareness, empathy, and grief.

There are three genera within the Elephantidae family:

9.6.1.1. Genus: Loxodonta (African elephants)

This genus comprises two species: the African Bush Elephant (Loxodonta africana) and the African Forest Elephant (Loxodonta cyclotis). African elephants are the largest land animals alive today.

African Forest Elephant

9.6.1.2. Genus: Elephas (Asian elephants)

This genus consists of a single species: the Asian Elephant (Elephas maximus). They are smaller than African elephants and have smaller ears and a single-humped back.

Asian Elephant (Elephas maximus)

Both African and Asian elephants are threatened by habitat loss, poaching, and human-elephant conflict. Conservation efforts are underway to protect these magnificent animals. Elephants are social animals that live in herds led by females. They are intelligent creatures with complex social behaviors and good memories. Unfortunately, all elephant species are threatened by habitat loss and poaching for their tusks.

9.6.1.3. Genus: Mammuthus (Mammoths)Extinct

Mammoths were elephant relatives that roamed the Earth during the Pleistocene epoch (2.6 million to 11,700 years ago) and became extinct relatively recently. There are several mammoth species identified, including the Woolly Mammoth (Mammuthus primigenius).

Woolly Mammoth (Mammuthus primigenius).

9.6.2. Challenges of Elephant Anaesthesia

Anesthetizing elephants, members of the Elephantidae family, is a complex and delicate undertaking. Due to their immense size, unique physiology, and social nature, special considerations are needed to ensure their safety and well-being during procedures. Here's a breakdown of anaesthesia for these gentle giants:

- ***Massive size:*** Delivering the correct anaesthetic dosage is crucial, and their size makes it challenging to calculate the right amount.
- ***Respiratory depression:*** Elephants are susceptible to breathing difficulties under anaesthesia, requiring close monitoring and potential ventilator support.
- ***Stress susceptibility:*** The capture and handling process can be stressful for elephants, potentially affecting the anaesthetic response.
- ***Heat regulation:*** They can overheat easily due to their large size and minimal sweat glands. Maintaining their body temperature during anaesthesia is vital.

9.6.3. Anaesthetic Techniques for Elephants

- ***Field Immobilization:***Anaesthesia for elephants is typically performed in the field, near their habitat, to minimize stress.

- ***Remote injection:*** Drugs are often delivered via darts fired from a distance to minimize stress and ensure safety for the animal and the team.
- ***Combination Anaesthesia:*** A combination of drugs is used for a safe and effective procedure:
  - ***Opioids:*** Powerful opioids are used for initial sedation, but due to their sensitivity, the dosage needs to be carefully calculated.
  - ***Alpha-2 adrenoreceptor agonists:*** Drugs like detomidine or medetomidine provide sedation and pain relief.
  - ***Muscle relaxants:*** Neuromuscular blocking agents might be used to facilitate specific procedures.
- ***Monitoring:*** Throughout the procedure, vital signs like heart rate, respiration, oxygen levels, and body temperature are continuously monitored.

9.6.4. Anaesthetic Considerations

- Pre-anesthetic assessment: A thorough health check is crucial to identify any underlying conditions that could complicate anesthesia.
- Stress management: Techniques to minimize stress during capture and handling are essential for a smooth procedure.
- Recovery: Elephants require a quiet and controlled environment to recover from anesthesia, with close supervision until they are fully awake and stable.
- Minimizing stress: Techniques to minimize capture and handling stress are essential for the safety and well-being of the elephant.
- Ethical considerations: Anesthesia should only be used for essential medical procedures and not for entertainment purposes.
- Due to the complexities involved, elephant anesthesia should only be attempted by a team with the necessary expertise and resources. The focus should always be on minimizing risks and ensuring the elephant's safety throughout the entire process.
- Elephantidae require a carefully monitored recovery period until they are fully awake and stable. This might involve providing them with support to stand and ensuring they maintain a proper body temperature.

Anesthetizing an elephant is a specialized procedure requiring a team of experts, careful planning, and specialized equipment. It's a high-risk but sometimes necessary procedure for veterinary care and conservation efforts.

The anaesthetic team should include: Veterinarians specializing in large animal anesthesia; specialists in elephant handling and behavior, support staff trained in monitoring and emergency procedures.

9.6.5. Immobilization of Elephant

The Asian elephant *(Elephus maximus)* was successfully immobilized using etorphine HCl (M-99). The 8 mg of Cyprenorphine (M-285) reversed the immobilizing effects almost immediately. The animal was immobilized using completely 2 ml capacity projectile syringes fitted with special elephant needles of 7 cm length with a 5 mm outside diameter and a barb 4.5 cm from the tip. Cyprenorphire 25 mg was administered intravenously in the marginal ear vein (Jainudeen, 1970).

Etorphine HCl offers several advantages.

The dose could be accurately computed since a record of the animal's body weight, height, and condition was available.

The animal does not react adversely to the impart of the syringe and needle and, therefore, could be immobilized at a site where it could be secured.

Its use precludes the unnecessary death of aggressive animals.

Etorphine HCl has been used to castrate Asian elephants at a rate of 6 mg administered intramuscularly into the semi-membronosis muscle of the left hind limb. Anesthesia was maintained with intermittent administration of 1 mg of etorphine into the ear vein (Fowler and Harter, 1973).

Xylazine at the rate of 100-300 mg of a 10% solution has been used satisfactorily to sedate the elephant in the standing position. The induction time ranged from 10±4 to 20±4 minutes. The duration of effect was 60±8 to 100±15 minutes. Complete recovery ranged from 360±31 to 540±21 minutes. Repeated injections at 3-4 day intervals have no adverse effects (Bongso, 1979).

Heavy xylazine sedation in an adult Asian elephant has been antagonized successfully by intravenous administration of yohimbine and 4-aminopyridine (Schmidt, 1983).

Yohimbine was found to be an effective antagonist to the sedation and immobilization of an elephant induced by a combination of xylazine and ketamine. Immobilized elephants were able to stand within 5 minutes of the administration of the drug whereas, non-antagonized elephants standing time was 11.6±6.9 minutes (Elloitt *et al.*, 1985).

Flaxedil (Gallamine triethiodide) (1600 mg) was administered intramuscularly in the gluteal muscles of an elephant. Five minutes later, interval sodium (4 grams in 160 ml of distilled water) was administered slowly to the ear vein.

The duration of the anesthesia was 40 minutes. The recovery time was 80 minutes after the induction of anesthesia (Muraleedharan *et al.*, 1979).

Recommendations

- Etorphine-acepromazine gives excellent results (1 mg/500 kg : 0.01-0.02 mg/kg administered intramuscularly).
- Antagonists: Diprenorphine (1.0 mg/kg body weight) should be given intravenously in an ear vein to reverse narcosis.
- Immobilized elephants should not be allowed to lie in sternal recumbency, as the pressure of the abdominal viscera on the diaphragm causes respiratory depression. They should always be pulled over onto their side (Green, 1982).
- The use of promazine tranquilizers causes relaxation or transient paralysis of the penis in male animals in this group (Wallach and Boever, 1983).

10

Chemical Restraint of Primates

10.1. Classification of Primates

The classification of primates is a complex topic that has been revised over time as our understanding of these fascinating creatures has evolved. Here's a breakdown of the current classification system:

10.1.1. Order: Primates

Primates are an order of mammals that includes lemurs, lorises, tarsiers, monkeys, apes, and humans. They are characterized by a number of shared features, including:

- ***Opposable thumbs:*** This allows primates to grasp objects and manipulate their environment.
- ***Forward-facing eyes:*** This provides primates with depth perception, which is important for arboreal locomotion and foraging.
- ***Enlarged brains:*** Primates have relatively large brains for their body size, which is associated with their intelligence and complex social behaviors.
- ***Reduced sense of smell:***Compared to other mammals, primates have a weaker sense of smell. This is thought to be due to their reliance on vision for foraging and social interaction.

The order Primates is divided into two suborders:

10.1.1.1. Suborder Strepsirrhini (wet-nosed primates)

This suborder includes lemurs, lorises, and pottos. Strepsirrhines have a moist rhinarium (wet nose) and a grooming claw on the second toe of their hind feet. They are generally nocturnal animals and tend to have smaller brains than haplorrhines.

Strepsirrhini primates

10.1.1.2. SuborderHaplorrhini (dry-nosed primates)

This suborder includes tarsiers, monkeys, apes, and humans. Haplorrhines have a dry rhinarium and lack a grooming claw. They are generally diurnal animals and tend to have larger brains than strepsirrhines.

Haplorrhini primates

The suborder Haplorrhini is further divided into two infraorders:

1. **Tarsiiformes:**This infraorder includes only one family, the Tarsiidae, which contains tarsiers. Tarsiers are small, nocturnal primates with large eyes, long hind legs, and a sticky pad on their third finger that they use for grooming.

Tarsiiformes primates

2. **Simiiformes (simians):** This infraorder includes monkeys, apes, and humans. Simiiformes are characterized by their relatively large brains, forward-facing eyes, and opposable thumbs on both their hands and feet. The infraorder Simiiformes is further divided into two parvorders:

   i) **Platyrrhini (New World monkeys):** This parvorder includes monkeys that are found in Central and South America. Platyrrhines have flat nostrils that are directed to the sides of their faces. They also have a prehensile tail, which they can use for grasping branches and objects.

Platyrrhini primates

   ii) **Catarrhini (Old World monkeys and apes):** This parvorder includes monkeys and apes that are found in Africa and Asia, as well as humans. Catarrhines have narrow nostrils that are directed downward. They also lack a prehensile tail (except for barbary macaques).The parvorder Catarrhini is further divided into three superfamilies:

a. **Cercopithecoidea (Old World monkeys):** Old World monkeys are primates in the family Cercopithecidae. Twenty-four genera and 138 species are recognized, making it the largest primate family. Old World monkey genera include baboons (genus *Papio*),redcolobus(genus *Piliocolobus*),and macaques (genus *Macaca*). Common names for other Old World monkeys include the talapoin, guenon, colobus, douc (douc langur, genus *Pygathrix*), vervet, gelada, mangabey (a group of genera), langur, mandrill, surili (*Presbytis*), patas, and proboscis monkey.

Cercopithecoidea primates

b. **Hominoidea (apes and humans):** This superfamily includes apes (gibbons, orangutans, gorillas, chimpanzees, and bonobos) and humans.

Hominoidea primates

c. **Indriidae (indris):** This superfamily includes a group of lemurs that are found in Madagascar. Indriids are characterized by their large size, loud vocalizations, and adaptations for leaping locomotion.

Indriidae primates

The classification of primates is an ongoing area of research, and new discoveries are constantly being made. However, the current system provides a framework for understanding the diversity and relationships of these amazing creatures.

10.2. Chemical Immobilization

Chemical restraint, also called chemical immobilization, is the use of drugs to immobilize primates for medical procedures, research, or transport. There are several reasons why chemical restraint might be necessary for primates. For example, a primate may need to be immobilized for a blood draw or other medical procedure that could be painful or stressful. Chemical restraint can also be used to transport primates safely, or to restrain them during research procedures.It is a common practice in primate research, but it is important to note that there are ethical concerns surrounding the use of primates in research. It involves administering drugs that produce a state of sedation or anaesthesia.

There are a number of different drugs that can be used for chemical restraint in primates. The most commonly used drugs are Telazol (a combination of tiletamine and zolazepam), ketamine, and medetomidine. The specific drug that is chosen will depend on the age, species, and health of the primate, as well as the specific procedure that is being performed.Chemical restraint can be a safe and effective way to immobilize primates, but it is important to use the correct drugs and to monitor the animal closely during the procedure.There are several different types of chemicals that can be used to restrain primates, including:

- ***Anesthetics:*** These drugs induce a state of unconsciousness, which allows for painful or invasive procedures to be performed without causing distress to the animal.

- ***Tranquilizers:*** These drugs produce a state of sedation, which makes the animal calmer and easier to handle.
- ***Muscle relaxants:*** These drugs can be used to relax the muscles, which can be helpful during certain procedures.

The specific type of chemical restraint that is used will depend on the species of primate, the age and health of the animal, and the procedure that is being performed.Here are some of the important considerations for chemical restraint inprimates:

- ***Anesthesia versus sedation:***Anesthesia is a state of unconsciousness in which the animal does not feel pain. Sedation is a state of drowsiness in which the animal may still feel pain. The type of chemical restraint that is chosen will depend on the specific procedure that is being performed.
- ***Drug selection:*** The specific drug that is chosen for chemical restraint will depend on the age, species, and health of the primate. It is important to use a drug that is safe and effective for the particular animal.
- ***Monitoring:*** Primates that are chemically restrained must be monitored closely throughout the procedure. This includes monitoring their heart rate, respiration, and temperature.
- ***Pain management:*** Even if an animal is anesthetized, it may still feel pain. It is important to use pain medication when appropriate.
- ***Recovery:*** Primates must be allowed to recover from chemical restraint in a quiet and comfortable environment.

Chemical restraint can be a safe and effective way to immobilize primates, but it is important to use the correct drugs at the correct dosage. Improper use of chemical restraint can lead to serious health problems, including respiratory depression, cardiac arrest, and even death.Chemical restraint should only be performed by a trained veterinarian or other qualified professional. When used properly, chemical restraint can be a safe and effective way to immobilize primates for medical procedures, research examinations, or transport.

Here are some of the important considerations for the chemical restraint of primates:

- The animal should be fasted for a period of time before the procedure to reduce the risk of regurgitation.
- The animal should be given a thorough physical examination before the procedure to ensure that it is healthy enough to undergo chemical restraint.
- The drugs should be administered by a trained veterinarian or other qualified professional.

- The animal should be monitored closely after the procedure to ensure that it recovers safely.
- Alternatives to chemical restraint are being developed, such as training primates to cooperate with procedures. However, chemical restraint is still a necessary tool in many cases.

10.3. Anesthetic techniques

It is always undesirable to handle conscious monkeys, as they not only inflict bites and scratches, but they are also potential carriers of viruses highly pathogenic to humans, such as tuberculosis, salmonellosis, and shigellosis. The drug should be soluble in a small volume thatcan be rapidly injected and rapidly absorbed from accessible intramuscular sites. The therapeutic index should be wide enough to allow dosages to be calculated based on an estimated weight. Finally, normal appetite and behavior should be reestablished quickly after recovery.

Several trials have been conductedusing narcotic and tranquilizing agents onvarious species of monkeys. Onbaboons', barbiturates likepentobarbitone, hexobarbitone, and amylbarbitone, and inhalational anesthetics like ether, N_2O, and chloroform, as well as other CNS depressants including chlorpromazine, paraldehyde a-chloralose, chloral hydrate, and morphine, have been assessed (Green, 1982) and concluded that the barbiturates were unsuitable for administration by intramuscular or oral routes. The response to intramuscular pentobarbitone was unpredictable. Paraldehyde may help in reducing pentobarbitone dosage, but the response is unpredictable. Chloral hydrate was not accepted orally. The α-chloralose produced heavy sedation at a dose of 200 mg/gm if taken in bait, but the clonic movements of hind limbs were noted. Morphine was ineffective, whether administered orally or intramuscularly. Chlorpromazine was considered valuable at dose rates up to 3.5 mg/kg of body weight given intramuscularly, although the slow attainment of peak effect (2-6h) was a disadvantage. Phencyclidine was the first of the dissociative anesthetic agents to be extensively used in sub-human primates. It was found that CNS depression was dose-dependent, ranging from sedation with early loss of the bite reflex through ataxia to cataleptic narcosis, and profound analgesia was observed. At threetimes the dose needed to induce cataleptic narcosis, the CNS was stimulated to produce convulsions, and finally, respiratory failure.

Clinical signs after intramuscular administration of Phencyclidine to subhuman primates

S.No.	Dosage mg/kg (I/M)	Observations
1	0.1	Face is normally alert, but eye blink slows, monkey repeatedly licks lips, and bite reflex decreases. Animal backs into corner of cage it approached
2	0.3	The face appears slightly anxious; the eyes blink noticeably slowly, and although tone in the jaw muscles has been retained, the biting reflex is completely lost. The monkey becomes increasingly ataxic, sits in a corner, and rocks to and fro on haunches. Movements of the eye ball (nystagmus) may be observed.
3	0.5	The facial expression appears sad. Eyelids partially close (ptosis). The animal can be safely picked up and handled.
4	1.0	The animal is usually immobilized yet apparently conscious. Attempts to mold the limbs resisted, and muscle tone was strong. Analgesia is usually well developed at this stage. Vertical and horizontal nystagmus are commonly exhibited.
5	3.0	A state of cataleptic stupor develops in which there is no response to pain or other stimuli. Eyelids are partially closed, and respiratory and cardiovascular function are near normal. Animals may be handled in less than 10 minute.
6	5.0	A state resembling light surgical anesthesia was developed within 6 minutes of the administration of the drug. The tongue protrudes, and profuse salivation is exhibited unless atropine is previously administered. The eyelids often blink. Nystagmus is common, and the pupil may dilate and contract in rapid succession. Analgesia is well developed, but muscle tone is still marked.
7	10.0	Tetanic spasms of the limbs progress to clonic extensor spasms.

James (1965) used phencyclidine in baboons (*Papio papio*). It hasa wide margin of safety, a rapid effect, is administered intramuscularly, and is especially useful for immobilizing baboons. The average dosage used is 1.2 mg/lb of body weight. Onset of action formost baboons took only 2-3 minutes. Recumbency occurred 2-19 minutes after administration.Edward (1965) evaluated the anesthetic properties of phencyclidine in simian primates. At high dose levels, involuntary movements were noticed, which resembled those associated with inadequate dosage.

Intramuscular administration of phencyclidine in Rhesus monkeys *(Macaca mulatta)* (Edward, 1965)

Average dose (mg/kg of body weight)	**Average time of onset (minutes)**	**Effect**	**Average time to recovery (hours)**
0.25	2	Calmness	½
0.5	2	Analgesia	1
0.15	3	Catalepsy	1 ½
1.00	5	Surgical anesthesia	2
1.5-5.0	5	Surgical anesthesia	2-4
5.0 and above	6	Clonic spasms progress to intermittent convulsions.	4-6

Oral administration of phencyclidine in Rhesus monkeys *(Macaca mulatta)* (Edward, 1965)

Average dose (mg/kg of body weight)	**Average time of onset (minutes)**	**Effect**	**Average time to recovery (hours)**
0.75	55-65	Calmness to analgesia	2
1.0	55	Analgesia	2
2.0-5.0	45-55	Surgical anesthesia	4

Immediately after injection with phencyclidine, the animal will assume its original, usually aggressive attitude. In approximately two minutes, he becomes quicker and gradually assumes a sitting position, slowly losing his grip on the cage or other surrounding objects. Within 4-5 minutes, catalepsy is ensured. The induction is quite smooth, without an excitatory stage. Muscular response, i.e., relaxation adequate for major surgery, was not obtained regardless of the dose. Ketamine is considered the agent of choice for chemical restraint inmost sub-human primates (Edwards, 1965).

Beck (1972) usedketamine in 24 species of primates. He found that, in general, larger species tended to require smaller dosages on amg/kg basis. Very young animals tended to require a higher dosage. Induction was rapid, and the duration of anesthesia varied dependingupon initial dosages and supplemental dosages. The emergence of recovery from anesthesia occurred between 1 and 3 hours after the initial injection. The medium period of duration was within the range 20-55 minutes. There was no evidence that ketamine was in any way contraindicated in pregnant or breeding females. Excessive salivation was the second-most common side effect. The first is vomiting, whichwas easily controlled with atropine given at arate of 0.01-0.05 mg/kg of body weight.

Suggested initial dosages of ketamine for Primates (Beck, 1972)

Species	**For Chemical restraint (mg/kg of body weight)**	**For Surgical anesthesia (mg/kg of body weight)**
Owl monkeyor Night monkey*(Aotustrivirgatus)*	10.0-12.0	20.0-25.0
Capuchin or White face monkey *(Cebus capuchin)*	13.0-15.0	25.0-30.0
White eyelid Mangabey monkey *(Cercocebuslophocebus kipunji)*	5.0-7.0	10.0-15.0
Grivets (*Chlorocebus aethiops)*	10.0-12.0	25.0-30.0
Patas monkey or Hussar monkey *(Erythrocebuspatas*)	3.0-5.0	5.0-7.5
Gorilla *(Gorilla gorilla)*	7.0-10.0	12.0-15.0
White Handed Gibbon or Lar Gibbon *(Hyalobates lar)*	5.0-10.0	10.0-12.0
Lemur catta (ring tailed lemur)	7.0-10.0	10.0-12.0
Macaca fascicularis(Crab eating macaque)	12.0-15.0	20.0-25.0
Macaca fuscata (Japanese macaque or snow monkey)	5.0	
Macaca mulatta (Rhesus monkey)	5.0-10.0	20.0-25.0
Macaca nemstrina (Pig tailed macaque)	5.0-7.5	20.0-25.0
Macaca radiate (Bonnet macaque)	12.5-15.0	25.0-30.0
Macaca sitenus(lion-tailed macaque)	10.0-12.0	
Macaca speliosaarctoides (Stamp tailed macaque)	5.0-7.5	20.0-25.0
Miopithecus talapoin (Mangrove monkey)	5.0-7.5	10.0-12.5
Pantroglodytes(Common chimpanzee)	5.0-7.5	10.0-15.0
Papio Anubis (Olive baboon)	5.0-7.5	10.0-15.0
Papio cymocephalus (Yellow baboon)	5.0-7.5	7.5-10.0
Papio papio(Guinea baboon)	10.0-12.0	
Pango pygameus (Orangutan)	5.0-7.5	12.5-15.0
Presbyris entellus (Entellus langur	2.0-3.0	3.0-5.0
Saimiri sciureus(Squirrel monkey)	2.0-15.0	25.0-30.0
*Symphalangussyndactylus*Siamang)	5.0-7.0	7.0-10.0

Another dissociative agent,tiletamine, has been assessed for clinical use alone in sub-human primates (Bree, 1972) and in combination withzolazepam in dogs and primates. The onset of action is rapid after tiletamine injection; it has a wide margin of safety for routine and surgical procedures alone or in combination with pentobarbitone in the dose range of 2-6 mg/kg intramuscularly. And 6-17 mg/kg orally in orange juice. However, an analysis of their results suggests that the combination has no major advantages over ketamine, especially if ketamine is given together with xylazine or diazepam. Neuroleptoanalgesia with a combination of fentanyl and droperidol has been relatively successful in a wide range of primates. Dose rates are 0.02 mg/kg fentanyl with 2.0 mg/kg droperidolintramuscularin apes and 0.04 mg/kg fentanyl with 3.0 mg/kg droperidolintramuscularin monkeys. Produced sufficient analgesia for 30-60 minutes of surgery. This combinationhas a qualitatively similar pharmacological effect in dogs and primates. Primates did not appear to respond to auditory stimuli (Field, 1966). The respiratory depressant effect is easily reversed by administering nalorphine (Field, 1966). Etorphine is also effective at adose rate of 0.005 mg/kg; pethidine (10 mg/kg); andthiambutene (10 mg/kg).The peak effect was obtained at 15 minute and lasted 60-120 min (Lumb and Jones, 1984).

Steroid combination alphaxalone-alphadalone at adose rate of 12 mg/kg sedated monkeys, higher doses up to 60 mg/kg intramuscularly produced light surgical anesthesia for up to 3 hours. The recommended technique for full surgical anaesthesia was to give 10 mg/kg i.m., wait 5-7 minutes for sedation to develop, and then give a further 6-12 mg/kg i.v. to effect.Desirable features includea wide margin of safety that can be givenintravenously or intramuscularly, smooth and rapid induction, and good muscular relaxation. Respiration is not significantly depressed except at an unnecessarily high dose level administered intravenously. Recovery period is short and uncomplicated animals rapidly gain full appetite. Steroids are relatively insoluble, and the volume is inconvenient for rapid intramuscularinjection in large primates. Steroids must not be used after barbiturates. It was recommended as the agent of choice for marmosets at doses up to 18 mg/kg intramuscularly.

Inhalational anesthesia without previous chemical immobilization can be achieved by putting primate into an induction box into which anesthetic gases may be piped. N_2O, after the animal was unconscious, was removed from the box maintained by using an ether-air mixture,halothane-O_2 mixture, and methoxyflurane.

10.4. Recommendedanesthetic technique

Small species, such as marmosets and squirrel monkeys, should be fasted for 6 hours prior to anaesthesia. Adult macaques and large animals should have no food for 24hours. All animals should be allowed free access to water until the first drugs are administered.

Intramuscular injection is best made either into the belly of the deltoid muscle over the shoulder, the lateral aspect of the anterior thigh (quadriceps groups of muscles), or the lateral aspect of the poster for thigh (gastronomies muscle). Intravenous injection can be given at the brachial (radial) vein in the forearm, the recurrent tarsal vein on the posterolateral aspect of the hindlimb, or the femoral vein. Atropine at 0.05 mg/kg subcutaneously or intramuscularly should be routinely administered as early as possible in the anesthetic sequence.

Ketamine is the agent of choice for the restraint of all primates except marmosets and squirrel monkeys. Dose level: 5-25 mg/kg body weight, and 4.0 mg should not be exceeded. Onset of effect: 2-6 minutes. Duration of action: 20-60 minutes. Recovery: 40-140 minutes. A few responses are less desirable: tongue protrusion andexcessive salivation may contribute toair-way obstruction, unless a monkey has been given atropine, laryngeal and pharyngeal reflexes are only partially preserved and protection of the air-way is incomplete interfere with endotracheal intubation, anesthesia has to be deepened or muscle relaxants administered before a tube can be passed. Responses mentioned earlier Alphaxolone-alphadolone is the agent of choice in marmosets and squirrel monkeys

12-18 mg/kg onset of action within 10 minutes

Duration of action:10 minutes, recovery complete in 45 minutes.

Other mentioned anesthetic agents can also be used.While recovering, attention to heat conservation, air-way patency, and fluid balance are of prime importance during post-operative care.Recovery is allowed to occur at an ambient temperature of 30°C.

References

Addison EM, Kolenosky GB (1979) Use of ketamine HCl and xylazine HCl to immobilize black bears (Ursus americanus). Journal of Wild Life Diseases, 15: 253-258.

Alford BT, Burkhart RL, Johnson WP (1974) Etorphine and diprenorphine as immobilizing and reversing agents in captive and free-ranging mammals. Journal of American Veterinary Medical Association, 164(7): 702-705.

Allen JL and Oosterhuis JE (1986) Effect of tolazoline on xylazine-ketamine induced anesthesia in turkey vultures. Journal of American Veterinary Medical Association, 189(9): 1011-1012.

Arnemo JM, Evans AL, Fahlman A, Caulkett N (2014) Field emergencies and complications. In West G, Heard D, Caulkett N (eds) Zoo Animal and Wildlife Immobilization and Anesthesia. (2nd edn) Ames, Iowa, USA, Wiley Blackwell Publishing, 139-148.

Arnemo JM, Soli E (1992) Immobilization of mink (Mustela vison) with medetomidine-ketamine and remobilization with atipamezole. Veterinary Research Communications, 16: 281-292.

Arora BM (1996) Chemical capture of escaped wild pigs (Sus scrofa cristatus) using Immobilon 11(3): 1-2, Zoos' Print, In: A Compendium of publications from Indian Zoos, Vol II, 2000 (Health and Disease Management), Compilers Acharjyo LN and Patnaik SK, Indian Zoo Directors' Association and Central Zoo Authority, Page 243-244.

Atkinson M, Kock M, Meltzer D (2012) Principles of chemical and physical restraint of wild animals. In Kock M, Burroughs R (eds) Chemical and Physical Restraint of Wild Animals-A Training and Field Manual for African Species. (2nd edn) International Wildlife Veterinary Services, 103-130.

Barnard SM, Dobbs JS (1980) A hand made blow gun dart: Its preparation and application in a zoological park. Journal of American Veterinary Medical Association, 177: 951.

Bauditz R (1972) Sedation, immobilization and anesthesia with Rompun in captive and free living wild animals. Veterinary Medicine Review, 3/4: 204-226.

Beck CC, Dresner AJ (1972) Vetalar (Ketamine HCl) a cataleptic anesthetic agent for primate species. Veterinary Medicine and small Animal Clinician, 67: 1082-1084.

Beck CC, Ott BS (1971) Evaluation of Vetalar (Ketamine HCl)-A unique feline anesthetic. Veterinary Medicine and Small Animal Clinician, 66: 993-996.

Bednarski RM, Ludders JW, LeBlanc PH, Pickett JP, Sedgwick CJ (1985) Isoflurane-nitrous oxide anesthesia in an Andean condor (Vultur gryphus), Journal of the American Veterinary Medical Association 187(11): 1209-1210.

Beever WJ, Wright W (1975). Use of ketamine for restraint and anesthesia of birds. Veterinary Medicine and small Animal Clinician, 70: 86-88.

Bennett RA (1996) Anesthesia. Reptile Medicine and Surgery, Mader R(eds), W B Saunders, London, page241-247.

Bennett RA (1998) Reptile Anesthesia. Seminars in Avian and Exotic Pet Medicine, 7(1): 30-40.

Bogan JA, Mackenzie G, Snow DH (1978) An evaluation of tranquilization for use with etorphine as neurolept analgesic agents in the horse. Veterinary Record, 103: 471-472.

Bonal BS, Barthakur T, Thakuria DB, Sarmah KK, Baruah M (1994) Shifting AfroAsiatic lion by chemical immobilization in Assam state zoo, Guwahati. 9(9): 13. Zoos' Print, In: A compendium of publications from Indian Zoos, Vol II, 2000 (Health and Disease Management), Compilers Acharjyo LN, Patnaik SK, Indian Zoo Directors' Association and Central Zoo Authority, Page 242.

Bongso TA (1979) Sedation of the Asian elephant with xylazine. Veterinary Record, 105: 442.

Bree MM (1972) Clinical evaluation of tiletamine as an anesthetic in six non-human primate species. Journal of American Veterinary Medical Association, 161(6): 693-695.

Bristol DG, Smith J, Silberman MS (1984) Acepromazine and etorphine for prolonged anesthesia of a zebra. Journal of the American Veterinary Medical Association, 185(11): 1439-1440.

Brodbelt DC, Blissitt KJ, Hammond RA, Neath PJ, Young LE, Pfeiffer DU, Wood JLN (2008) The risk of death: The confidential enquiry into perioperative small animal fatalities. Veterinary Anaesthesia and Analgesia, 35: 365-373.

Bronstand A (2022) A good anesthesia practice for fish and other aquatics. Biology, 11: 1355.

Burke JJ, Wall BE (1970) Anesthetic death in cobras (Naja naja and Ophiophagus hannah) with methoxyflurane. Journal of American Veterinary Medical Association 157(5): 620-621.

Burroughs R, Meltzer D, Morkel P (2012) Applied Pharmacology. In Kock M, Burroughs R, Chemical and physical restraint of wild animals-A training and field manual for African species (2nd edn), International Wildlife Veterinary Services, 53-80.

Bush M, Grobler D, Raath J (2002) The Art and Science of Giraffe (Giraffa camilopardalis) Immobilization/Anesthesia, B0169.0102.

Bush M, Ensley P, Mehren K, Rapley W (1976) Immobilization of giraffes with xylazine and etorphine hydrochloride. Journal of the American Veterinary Medical Association 169(9): 884-885.

Bush M, Smeller JM (1978) Blood collection and injection techniques in snakes. Veterinary Medicine and small Animal Clinician, 73: 211-214.

Calderwood HW (1971) Anesthesia for Reptiles Journal of the American Veterinary Medical Association, 159: 1618-1625.

Capture and Immobilization of Free-Ranging Edentates (https://www.scribd.com/document/64291435/History-of-Entomology) (This PDF resource provides detailed information on edentate restraint techniques).

Carrasco S, Sumano H, Navohro-Fierro R (1984) The use of lidocaine-sodium bicarbonate as an anesthetic in fish. Aquaculture, 41: 161-163.

Cattet M R, Obbard ME (2010) Use of hyaluronidase to improve chemical immobilization of free-ranging polar bears (Ursus maritimus). Journal of Wildlife Diseases, 46(1): 246-250.

Caulkett NA, Arnemo JM (2007) Chemical Immobilization of Free-Ranging Terrestrial Mammals. In Tranquilli WJ, Thurmon JC, Grimm KA (eds), Lumb & Jones' Veterinary Anesthesia and Analgesia (4th edn). Ames, Iowa, USA, Blackwell, 807-832.

Caulkett N, Shury T (2014) Human safety during wildlife capture. In West G, Heard D, Caulkett N (eds) Zoo Animal and Wildlife Immobilization and Anesthesia. (2nd edn) Ames, Iowa, USA, Wiley Blackwell Publishing, 181-190.

Chamers GA, Barrett MW (1977) Capture myopathy in prong horns in Alberta Canada. Journal of the American Veterinary Medical Association, 171: 918-923.

Clarke KW, Trim CM, Hall LW (2014) Veterinary Anesthesia. (11th edn) Elsevier Saunders, St. Louis, MO, 10-244.

Comparison of two injectable anaesthetic protocols in Egyptian fruit bats (Rousettus aegyptiacus) undergoing gonadectomy [This research paper discusses anaesthetic options for bats undergoing procedures]

Cooper JE (1974) Ketamine HCl as an anesthetic for East African Reptiles. Veterinary Record, 95: 37-41.

Cooper JE, Jackson OF (1981) Diseases of Reptilian, Volume I and II, Academic Press, A subsidiary of Harcourt Bruce Jovanovich, Publishers, New York.

Cracknell J (2014) Airway Management. In West G, Heard D, Caulkett N (eds) Zoo Animal and Wildlife Immobilization and Anesthesia. (2nd edn), Ames, Iowa, USA, Wiley Blackwell Publishing, 53-64.

Custer RS, Bush M (1980) Physiologic and acid base measures of gopher snakes during ketamine or halothane-Nitrous oxide anesthesia. Journal of the American Veterinary Medical Association, 177(9): 870-874.

Custer R, Kramer L, Kennedy S (1977) Hematologic effects of xylazine when used for restraint of Bactrian camels. Journal of the American Veterinary Medical Association, 171: 899-901.

Daniel M, Long CM (1972) The effect of an etorphine/acepromazine mixture on the heart rate and blood pressure of the Horse. Veterinary Record, 90: 336-339.

David A, Jessup Clark WE, Karen R and others (1985) Immobilization of free-ranging desert bighorn sheep, tole, elk, and wild horses, using carefentanyl and xylazine. Reversal with naloxone, diprenorphine and yohimbine. Journal of the American Veterinary Medical Association, 187: 1253-1254.

David CT (1970) A simple technique for restraining birds. Veterinary Medicine and Small Animal Clinician, 65: 764.

David RC, Raymond J (1972) Effect of certain anesthetic agents on mallard ducks. Journal of the American Veterinary Medical Association, 161(6): 624-633.

De Vos V (1978) Immobilization of free ranging wild animals using new drug. Veterinary Record, 103: 64-68.

Edward C Ramsay (2020) Immobilization and Anaesthesia of Lions, Tigers and Leopards. Online International Conference: WILDCON-2020, Insights into wildlife conflicts, rescue and rehabilitation: Challenges and opportunities for conservation and 14th Annual convention of Association of Indian Zoo and Wild Life Veterinarians held at Maharashtra Animal and Fishery Sciences University, Nagpur, India from 18-20 December, 2020, WHM-LP2, page 100-105.

Edward CM, Baker HJ (1965) Phencyclidine for analgesia and anesthesia in Simian primates. Journal of the American Veterinary Medical Association, 147(10): 1068-1072.

Engelhardt FR (1977) Immobilization of harp seals (Phoca groenlandica) by intravenous injection of ketamine. Comp Biochemistry and Physiology C Comp Pharmacology, 56(2): 75-76 [doi: 10.1016/0306-4492(77)90016-8].

Erhardt W, Haberstroh J (2004) Species specific anesthesia. In Erhardt W, Henke J, Haberstroh J (eds) Anesthesia and analgesia in pets as well as birds, amphibians and fish, Stuttgart, Germany, page 425-663.

Fahlman A (2008) Advances in Wildlife Immobilization and Anesthesia: Clinical and Physiological Evaluation in Selected Species. Doctoral Thesis, Uppsala, Swedish University of Agricultural Sciences, 13-58.

Ferreira TH, Mans C (2022) Sedation and anesthesia of lizards. Veterinary Clinics of North American Exotic Animal Practice, 25: 73–95. [https://doi.org/10.1016/j.cvex.2021.08.002]

Field WE, Yelnosky J, Mundy J, Mitchell J (1966) Use of droperidol and fentanyl for analgesia and sedation in primates. Journal of the American Veterinary Medical Association, 149: 896-901.

Flecknell PA, Thomas AA (2015) Comparative anesthesia and analgesia of laboratory animals. In Veterinary Anesthesia and Analgesia (5th edn) Grimm KA et al (eds) John Wiley & Sons, Inc. Iowa, USA.

Fleming GJ (2001) Crocodilian anesthesia. Veterinary Clinics of North American Exotic Animal Practice, 4(1): 119-45. [https://doi.org/10.1016/s1094-9194(17)30054-3]

Fleming J (2014) Crocodilians (crocodiles, alligators, caiman, and gharial). In West G, Heard D, Caulkett N (eds),Zoo Animal and Wildlife Immobilization and Anesthesia (2nd edn) Blackwell Publishing, Ames, IA, USA, page325–336.

Fletcher J (1974) Hypersensitivity of an isolated population of Red deer (Cervus elaphus) to xylazine. Veterinary Record, 94: 85-86.

Foerster CR (1998) Ecología de la danta Centroamericana (Tapirus bairdii) en un bosque húmedo tropical de Costa Rica. MS Thesis. Universidad Nacional de Costa Rica, Costa Rica.

Foerster SH, Bailey JE and others (2000) Butorphanol/Xylazine/Ketamine Anaesthetic protocol for the immobilization of free-ranging Baird's tapirs in Costa Rica. Journal of Wildlife and Diseases, 36(2): 335-341.

Fowler ME (1974) Restraint and anesthesia in zoo animal practice. Journal of the American Veterinary Medical Association, 164: 706-711.

Fowler ME (2008) Restraint and handling of wild and domestic animals (3th edn), Iowa: Blackwell Publishing.

Fowler ME, Hart R (1973) Castration of an Asian elephant using etorphine anesthesia. Journal of the American Veterinary Medical Association, 163: 539-543.

Frye F (1991) Biomedical and Surgical Aspects of Captive Reptile Husbandry. Krieger, Malabar, FL.

Gachen G, Quse V, Falzone M, González Ciccia P (2011) Effective drug combinations for collecting samples in lowland tapirs (Tapirus terrestris). In Proceedings of the Fifth International Tapir Symposium, page32.

Girling SJ (2013) Veterinary Nursing of Exotic Pets, Second Edition, Edited by Simon J. Girling, John Wiley & Sons, Ltd Published 2013 by John Wiley & Sons, Ltd, page272-285. [https://doi.org/10.1002/9781118782941.ch19]

George PO, Rajakutty K (1990) Translocation of a stripped (Hyaena hyaena) at Trichur Zoo 5(10): Zoos' Print, In A Compendium of publications from Indian Zoos, Vol II, 2000 (Health and Disease Management), Compilers Acharjyo LN, Patnaik SK. Indian Zoo Directors' Association and Central Zoo Authority, page 228.

George PO, Rajankutty K, Cheeran JV (1990) General anesthesia in Lion tailed monkey (Macaca silenus) using Droperidol-Fentanyl mixture (Innovar-Vet) 5(5): 13. Zoos' Print, In A Compendium of publications from Indian Zoos, Vol II, 2000 (Health and Disease Management), Compilers Acharjyo L.N. and Patnaik S.K., Indian Zoo Directors' Association and Central Zoo Authority, page 268.

Geraci JR (1973) An appraisal of ketamine as an immobilizing agent in wild and captive pinnipeds. Journal of the American Veterinary Medical Association, 163(6): 574-577.

Gilderhus PA (1991) Benzocaine as an anesthetic for striped bass. Progressive fish culturist, 53(2): 105-107.

Gilderhus PA, Marking LL (1987. Comparative efficacy of 16 anesthetic chemicals in rainbow trout. North American Journal of Fisheries Management, 7: 288-292.

Glenn JL, Straight R, Snyder CC (1972) Clinical use of ketamine HCl as an anesthetic agent for snakes. American Journal of Veterinary Research, 33: 1901-1903.

Goodman LS, Gilman AG, Gilman A (1980) The pharmacological basis of the therapeutics (6th edn) Macmillan Publishing Co. Inc, New York.

Goodman G, Hedley J, Meredith A (2013) Field techniques in zoo and wildlife conservation work. Journal of Exotic Pet Medicine, 22(1): 58-64.

Goodrich JM, Kerley LL, Schleyer BO, Miquelle DG, Quigley KS, Smirnov YN, Nikolaev IG, Quigley HB, Hornocker MG (2001) Capture and chemical anesthesia of Amur (Siberian) tigers. Wildlife Society Bulletin, 29(2): 533-542.

Gopal R (1991) Chemical immobilization of hyena (Hyaena hyaena) 6(5): Zoos' Print, In A Compendium of publications from Indian Zoos, Vol II, 2000 (Health and Disease Management), Compilers Acharjyo LN, Patnaik SK, Indian Zoo Directors' Association and Central Zoo Authority, page 227.

Graham M, Iwama GK (1990) The physiologic effects of the anesthetic ketamine hydrochloride on two salmonid species. Aquaculture, 90(3-4): 323-331.

Green CJ (1978) Anesthetizing ferrets. Veterinary Record, 102: 269.

Greenaway JB, Partlow GD, Gonsholt NL, Fisher KR (2001) Anatomy of the lumbosacral spinal cord in rabbits. Journal of American Animal Hospital Association, 37: 27-34.

Greenberg S (1966) The use of combined inhalation anesthesia in laboratory animal surgery. Journal of the American Veterinary Medical Association, 149(7): 935-937.

Guidelines for anesthesia and analgesia in rats. Laboratory animal resources guidelines. Indiana University, Bloomington

Hadlow WJ, Vanderwall K, Young E (1974) Possible therapy for capture myopathy in capture wild animals. Nature, 247: 577.

Haigh JC, Hopt HC (1976) The blowgun in veterinary practice: Its use and preparation. Journal of the American Veterinary Medical Association, 169: 881.

Haigh JC, Lee LJ, Schweinsburg RE (1983) Immobilization of polar bear with carfentanil. Journal of Wildlife Diseases, 19: 140-144.

Hallett DL, John DR, Paddleford RR (1979) Immobilization of coyotes with ketamine and propiomazine. Journal of the American Veterinary Medical Association, 179(9): 1007-1008.

Harms CA, Bakal RS (1995) Techniques in fish anesthesia. Journal of Small Exotic Animal Medicine. 3:19-25.

Harthoorn AM (1966) Restraint of undomesticated animals. Journal of the American Veterinary Medical Association, 149: 875-880.

Harthoorn AM (1974) A relationship between balance and capture myopathy in zebra (Equus burchelli) and an apparent therapy. Veterinary Record, 95: 337-342.

Hernandez S (2014) Chemical immobilization of wild animals. In Clarke KW, Trim CM, Hall L W (eds) Veterinary Anesthesia. (11th edn) Saunders Elsevier, 571-584.

Hernández-Divers SM, Bailey JE, Aguilar R, Loria DL, Foerster CR (1998) Cardiopulmonary effects and utility of a butorphanol-xylazine-ketamine anesthetic protocol for immobilization of free-ranging Baird's tapirs (Tapirus bairdii) in Costa Rica. In Proceedings of the American Association of Zoo Veterinarians (AAZV).

Hernández-Divers SM, Bailey JE, Aguilar R, Loria DL, Foerster CR (2000) Butorphanol-xylazine-ketamine immobilization of free-ranging Baird's tapirs in Costa Rica. In Journal of Wildlife Diseases, 36(2): 335-341.

Hernández-Divers SM, Foerster CR (2001) Capture and immobilization of free-living Baird's tapirs (Tapirus bairdii) for an ecological study in Corcovado national park, Costa

Rica. In Zoological Restraint and Anesthesia, Heard D (Edn) International Veterinary Information Service, Ithaca, New York, USA.

Hofmeyr JM (1981) The use of haloperidol as a long-acting neuroleptic in game capture operations. Journal of the South African Veterinary Association, 52(4): 273-282.

Hofmeyr M, Fivaz B, Meltzer D (2012) Ancillary treatments in wildlife capture and care. In Kock M, Burroughs R (eds) Chemical and physical restraint of wild animals-A training and field manual for African species. (2nd edn) International Wildlife Veterinary Services, 311-318.

Houston AH, Corlett JT (1967) Specimen weight and MS-222. Journal of the Fisheries Research Board of Canada, 33: 1402-1407.

Hrbst LH, Packer C, Seal US (1985) Immobilization of free-ranging African lions (Panthera leo) with a combination of xylazine HCl and ketamine HCl. Journal of Wild Life Diseases, 21(4): 401-404.

Hsu WH, Shulaw WP (1984) Effect of yohimbine on xylazine induced immobilization in white tailed deer (Odocoileus virginianus). Journal of the American Veterinary Medical Association, 185: 1301-1304.

Hsu WH, Bellin SI, Dellmann HD, Habil V, Hanson CE (1986) Xylazine-ketamine induced anesthesia in rats and it's antagonized by yohimbine. Journal of the American Veterinary Medical Association, 9: 1040-1043.

Hubbell GL (1965) Capture and restraint of zoo animals. Journal of the American Veterinary Medical Association, 147(10): 1044-1047.

Immobilization and Translocation Protocol for the Giraffe (Giraffa camelopardalis) in Kenya (2019). Veterinary Services Department, Division of Biodiversity, Research & Planning, Kenya Wildlife Service.

Immobilization of Collared Peccaries (Tayassu tajacu) and Feral Hogs (Sus scrofa) with Telazol® and Xylazine [link to research paper on immobilization of feral hogs and peccaries] (This research paper explores the use of Telazol-xylazine combination for immobilizing both wild swine and peccaries)

Isaza R (2014) Remote drug delivery. In West G, Heard D, Caulkett N (eds) Zoo Animal and Wildlife Immobilization and Anesthesia. (2nd edn) Ames, Iowa, USA, Wiley Blackwell Publishing, page155-170.

Jacques CN, Jenks JA, Deperno CS, Sievers JD, Grovenburg TW, Brinkman TJ, Swanson CC, Stillings BA (2009) Evaluating ungulate mortality associated with helicopter net-gun captures in the northern great plains. The Journal of Wildlife Management, 73(8): 1282-1291.

Jacobson ER (1983) Hematologic and serum chemical effects of a ketamine-xylazine combination when used for immobilizing springbok. Journal of the American Veterinary Medical Association, 183(11): 1260-1262.

Jacobson ER, Allen J, Martin H, Kollias GV (1985) Effect of Yohimbine on combined xylazine-ketamine induced sedation and immobilization in juvenile African elephants. Journal of the American Veterinary Medical Association, 187(11): 1195-1198.

Jainuddeen MR (1970) The use of etorphine HCl for restraint of a domesticated elephant (Elephus maximum). Journal of the American Veterinary Medical Association, 157(5): 624-626.

James FV (1965) Phencyclidine anesthesia in Baboons (Papio papio). Journal of the American Veterinary Medical Association 147(10): 1073-1074.

Janssen DL, Rideout, BA and others (1999) Tapir Medicine. In Fowler ME, Miller RE (eds) Zoo and Wild Animal Medicine: Current Therapy (4th edn), Philadelphia, WB Saunders Co, page562-568.

Janssen DL (2003) Tapiridae. In Fowler ME, Miller RE (eds) Zoo and Wild Animal Medicine Saunders, Saint Louis, MO, USA, page569-577.

Janssen DL, Rideout BA, Edwards MS (1999) Tapir Medicine. In Fowler ME, Miller RE (eds) Zoo and Wild Animal Medicine, WB Saunders Co, Philadelphia, page562-568.

Jessup DA, Clark WE, Gullet PA and others (1983) Immobilization of mule deer with ketamine and xylazine and reversal of immobilization with yohimbine. Journal of the American Veterinary Medical Association, 183: 1339-1340.

Jessup DA, Jones M, Kucera R (1985). Yohimbine antagonism to xylazine in free-ranging mule deer and desert bighorn sheep. Journal of the American Veterinary Medical Association, 187(11): 1251-1253.

Jolly DW, Madeley M, Thomas LE, Bucke D (1972) Anesthesia of fish. Veterinary Record, 91: 424-426.

Jones DM (1984) Physical and chemical methods of capturing deer. Veterinary Record, 114: 109-112.

Jones DM (1982) Veterinary Pharmacology and Therapeutics (5th edn) 1st Indian Reprint.

Josephe R, Karlskirnisson G, David JS (1981) A safe method for repeatedly immobilizing seals. Journal of the American Veterinary Medical Association, 179(11): 1192-1193.

K Chandrasekara Pillai (1992) Record of behavior of Asiatic Lion on being Immobilized at Nehru Zoological Park, Hyderabad 7(8): 18. Zoos' Print, In: A Compendium of publications from Indian Zoos, Vol II, 2000 (Health and Disease Management), Compilers Acharjyo LN, Patnaik SK, Indian Zoo Directors' Association and Central Zoo Authority, Page 239.

Kischinovsky M, Bertelsen MF (2011) Alfaxalone anesthesia in green iguanas and red-eared sliders. Proceedings of the European Association of Zoo and Wildlife Vets, Lisbon, Lisbon, Portugal, page113.

Kreeger TJ, Arnemo JM (2018) Handbook of Wildlife Chemical Immobilization, 5th Edition.

King JM, Brian CR, Bertram P and others (1978) Tiletamine and zolazepam for immobilization of wild lions and leopards. Journal of the American Veterinary Medical Association, 171(9): 894-898.

Kocan AA, Glenn BL, Thedford TR and others (1981) Effects of chemical immobilization on hematologic an serum chemical values in captive white tailed deer. Journal of the American Veterinary Medical Association, 179: 1153-1156.

Kocan AA, Thomas R, Thedford BL and others (1980) Myopathy associated with immobilization in captive white failed deer (Odocoileus virginianus). Journal of the American Veterinary Medical Association, 177(9): 879-881.

Kock M, Jessup D, Burroughs R (2012) Ballistics and projectile darting systems. In Kock M, Burroughs R (eds) Chemical and physical restraint of wild animals-A training and field manual for African species (2nd edn) International Wildlife Veterinary Services, page271-286.

Kock M, Morkel P (2012) Other restraint tools. In Kock M, Burroughs R (eds) Chemical and physical restraint of wild animals-A training and field manual for African species. (2nd edn) International Wildlife Veterinary Services, page305-310.

Kotwal P. C., Sharma B. C, Pandey D. K. (1991) Immobilization and radio-collaring of Golden Jackal (Canis aureus Linn). 6(11): 33-34 Zoos' Print, In: A Compendium of publications from Indian Zoos, Vol II, 2000 (Health and Disease Management), Compilers Acharjyo L.N. and Patnaik S.K., Indian Zoo Directors' Association and Central Zoo Authority, Page 232.

Krahwinkel DJ (1970) The use of Tiletamine HCl as an incapacitating agent for a lion. Journal of the American Veterinary Medical Association, 175(5): 622-623.

Kreeger TJ, Armstrong DL (2010) Tigers and Telazol: The unintended evolution of caution to contraindication. Journal of Wildlife Management,74(6): 1183-1185.

Kreeger TJ (1997) Handbook of Wildlife Chemical Immobilization. Published by International Wildlife Veterinary Services Inc., USA, page341

Kuehn G (1986) Tapiridae. In Fowler ME (eds) Zoo and Wild Animal Medicine (2nd edn) Philadelphia, WB Saunders Co, page931-934.

La Grange M (2012) Overview of capture methods and transport of wild animals. In Kock M, Burroughs R (eds) Chemical and physical restraint of wild animals-A training and field manual for African species. (2nd edn) International Wildlife Veterinary Services, page319-330.

Lamont LA, Grimm KA (2014) Clinical Pharmacology. In West G, Heard D, Caulkett N (eds) Zoo animal and wildlife immobilization and anesthesia. (2nd edn) Ames, Iowa, USA, Wiley Blackwell Publishing, page5-42.

Lawton MPC (1992) Anaesthesia. Manual of Reptiles, 1st edn, BSAVA, Cheltenham, Glasgow.

Lee J, Schweinsburg R, Kernan F and others (1981) Immobilization of polar bears with ketamine HCl and xylazine HCl. Journal of Wild Life Diseases, 17: 331-336.

Lewis RJ, Chalmers GA, Barret MW and others (1977) Captive myopathy in elk in Alberta, Canada, Journal of the American Veterinary Medical Association, 171: 927-932.

Love JA (1970) Use of fentanyl and droperidal in Guinea pigs, Lemmings, Ground Squirrels and Cats. Journal of the American Veterinary Medical Association, 157: 5, 675-677.

Lumb WV, Jones EW (1984) Veterinary Anesthesia. Second Edition, Lea and Febiger, Philadelphia.

Lyle A, Renecker A, Olsen DC (1985) Use of yohimbine and 4-amino pyridine to antagonize xylazine induced immobilization in North American Cervidae. Journal of the American Veterinary Medical Association, 187(11): 1199-1201.

Malley D (1997. Reptile anesthesia and the practicing veterinarian. Practice, 19: 351–368.

Mandelker L (1971) Practical techniques for administering inhalation anesthetics to birds. Veterinary Medicine and Small Animal Clinician, 66: 224-225.

Mandelker L (1972) Ketamine HCl as an anesthetic for Parakeets. Veterinary Medicine and Small Animal Clinician, 67: 55-56.

Mangini PR, Medici EP (1998) Utilizaçao de associaçao de cloridrato de medetomidina com cloridrato de tiletamina e zolazepam na contençao de Tapirus terrestris em vida livre: Relato de dois casos. In: Proceedings 22o Congresso Brasileiro e 4o Encontro Internacional da SZB. Salvador, Brazil, Sociedade de Zoologicos do Brasil.

Mangini PR (2007) Perissodactyla - Tapiridae (Anta). In Tratado de Animais Selvagens: Medicina Veterinária, Cubas ZS, Silva JCR, Catão-Dias JL (eds) Editora Roca, São Paulo, Brazil, page598-614.

Mangini PR, Velastin GO, Medici EP (2001) Protocols of chemical restraint used in 16 wild Tapirus terrestris. Archives of Veterinary Science, 6: 6-7.

Matola S, Cuaron, AD and others (1997) Status and Action Plan of Baird's Tapir (Tapirus bairdii). In Brooks DM, Bodmer RE (eds) Tapirs - Status Survey and Conservation Action Plan. IUCN/SSC Tapir Specialist Group. IUCN Gland, Switzerland and Cambridge, page29-45.

McFadden MS, Bennett RA, Reavill DR, Ragetly GR, Clark-Price SC (2011) Clinical and histologic effects of intracardiac administration of propofol for induction of anesthesia in ball pythons (Python regius). Journal of American Veterinary Medical Association, 239(6): 803-807. [doi:10.2460/javma.239.6.803. PMID: 21916763]

McTaggart, J., Kock, M., Hofmeyr, M. (2012). Helicopter and fixed-wing use in wildlife work. In Kock M Burroughs R (eds) Chemical and physical restraint of wild animals-A training and field manual for African species. (2nd edn) International Wildlife Veterinary Services, page131-142.

Mech, D.L., Delgiudice, G.D., Karns, P.D. et al. (1985). Yohimbine HCl as an antagonist to xylazine-ketamine immobilization of white tailed deer. J. Wild L. Dis., 21: 405-410.

Medici EP, Mangini PR and others (2001) Order Perissodactyla, Family Tapiridae (Tapirs). In Fowler ME, Cubas Z (eds) Biology, Medicine and Surgery of South American Wild Animals, Iowa, Iowa State University Press, page363-376.

Medici EP (2010) Assessing the Viability of Lowland Tapir Populations in a Fragmented Landscape. PhD Dissertation, Durrell Institute of Conservation and Ecology (DICE), University of Kent. Canterbury, UK.

Medici EP (2011) Family Tapiridae (TAPIRS). In Wilson DE, Mittermeier RA (eds) Handbook of the Mammals of the World-Volume 2: Hoofed Mammals, Lynx Edicions, Spain.

Medway W, McCormick JG, Ridgway SH, Crump JF (1970) Effects of prolonged Halothane anesthesia on some cetaceans. Journal of the American Veterinary Medical Association, 157(5): 576-582.

Meltzer D, Kock N (2012) Stress and Capture-related Death. In Kock M, Burroughs R (eds) Chemical and physical restraint of wild animals-a training and field manual for African species. (2nd edn) International Wildlife Veterinary Services, 81-102.

Miller MA, Buss P (2015) Rhinoceridae (Rhinoceroses). In Miller RE, Fowler ME (eds) Zoo and Wild Animal Medicine. (Volume 8) St Louis, Missouri, Elsevier Saunders, page538-546.

Montané J, Marco I, López-Olvera J, Perpiñán D, Manteca X, Lavín S (2003) Effects of acepromazine on capture stress in roe deer (Capreolus capreolus). Journal of Wildlife Diseases, 39(2): 375-386.

Morkel P, Kock M (2012) Safety and first aid in the field-weapons, drugs and animals, In Kock M, Burroughs R (eds) Chemical and physical restraint of wild animals-a training and field manual for African species. (2nd edn) International Wildlife Veterinary Services, page89-101.

Muir WWIII, Hubbell JA (1991) Equine Anaesthesia: Monitoring and Emergency Therapy, St Louis, Mosby Year Book (Available from amazon.com).

Muraleedharan KN, Chandrasekharan K, Jacob V and others (1979) General anesthesia in an elephant (Elephus maximus) Kerala Journal of Veterinary Sciences, 10: 197-200.

Mustafa S, Zlateva N (2018) Anesthesia, chemical restraint and pain management in snakes (serpentes): A review. Tradition and Modernity in Veterinary Medicine, 3 (4): 37–44.

Napier J, Armstrong DL (2014) Nondomestic Cattle. In West G, Heard D, Caulkett N (eds) Zoo Animal and Wildlife Immobilization and Anesthesia (2nd edn) Ames, Iowa, USA, Wiley Blackwell Publishing, page863-872.

Neiffer DL, Stamper MA (2009) Fish sedation, anesthesia, analgesia and euthanasia: Considerations, methods, and types of drugs. Institute of Laboratory Animal Research Journal, 50(4): 343-360.

Nielsen L (1999) Chemical Immobilization of Wild and Exotic Animals. Ames, Iowa, USA: Iowa State University Press.

Nunes LAV, Mangini PR, Ferreira JRV (2001) Order Perissodactyla, Family Tapiridae (Tapirs): Capture Methodology and Medicine. In Fowler ME, Cubas ZS (eds) Biology, Medicine and Surgery of South American Wild Animals, Ames, Iowa University Press, USA, page367-376.

Olsen YA, Einarsdottir IE, Nilssen KJ (1995) Metomidate anaesthesia in Atlantic salmon, Salmo salar, prevents plasma cortisol increase during stress. Aquaculture, 134(1-2): 155-168.

Olson ME, Vizzutti D, Mork DW, Cos AK (1993) Parasympatholytic effects of atropine sulfate and glycopyrrolate in rats and rabbits. Canadian Journal of Veterinary Research, 57: 254-258.

Olson ME, McCabe K (1986) Anesthesia in the Richardsons ground squirrel: Comparision of ketamine, ketamine-xylazine, droperidal fentanyl and sodium pentobarbital. Journal of the American Veterinary Medical Association, 189(9): 1035-1037.

Pafenaude RP (1979) Evaluation of fentanyl citrate, etorpine HCl, and Nalaxone HCl in captive polar Bears. Journal of the American Veterinary Medical Association, 175(9): 1006-1007.

Page CD (1986) Sloth bear immobilization with a ketamine-xylazine combination. Reversal with xylazine. Journal of the American Veterinary Medical Association, 189(9): 1050-1051.

Parás-García A, Foerster CR, Hernández-Divers SM, Loria DL (1996) Immobilization of free ranging Baird's Tapir (Tapirus bairdii). In Proceedings American Association of Zoo Veterinarians (AAZV). Pérez J (2013) Immobilization of Baird´s Tapir (Tapirus bairdii) using thiopental oxalate (A3080) in combination with xylazine and ketamine. In: Tapir Conservation- The Newsletter of the IUCN/SSC Tapir Specialist Group, 22(31): 15-19.

Pas A (2014) Gazelle and small antelope. In West G, Heard D, Caulkett N (eds) Zoo Animal and Wildlife Immobilization and Anesthesia (2nd edn) Ames, Iowa, USA, Wiley Blackwell Publishing, page843-856.

Paterson R, Samuels JX, Rybczynski N, Ryan MJ, Maddin HC (2020) The earliest mustelid in North America. Zoological Journal of the Linnean Society, 188(4) page1318–1339 [].

Paterson J (2014) Capture Myopathy. In West G, Heard D, Caulkett N (eds) Zoo Animal and Wildlife Immobilization and Anesthesia (2nd edn) Ames, Iowa, USA, Wiley Blackwell Publishing, page171-180.

Peshin JM, Nigam SC, Singh SC, Robinson BA (1980) Evaluation of xylazine in camels (Camelus dromedarous) Journal of the American Veterinary Medical Association, 177(9): 875-878.

Philo ML (1978) Evaluation of xylazine for chemical restraint of captive arctic wolves. Journal of the American Veterinary Medical Association, 173(9): 1163-1166.

Plumb DC (2008) Plumb's Veterinary Drug Handbook. Pharma Vet Inc, Stockholm, Wiley-Blackwel, Ames, USA [ISBN (Hardback): 978-1-118-91193-8].

Porter KR (1972) Herpetology, W B Saunders, Philadelphia, PA.

Quigley KS, Armstrong DL, Miquelle DG, Goodrich JM, Quigley HB (2001) Health evaluation of wild Siberian tigers (Panthera tigrisaltaica) and Amur leopards (Panthera pardus orientalis) in the Russian Far East. Proceedings of American Association Zoo Vet, page179-182.

Radcliffe RW, Morkel P (2014) Rhinoceroses. In West G, Heard D, Caulkett N (eds) Zoo Animal and Wildlife Immobilization and Anesthesia (2nd edn) Ames, Iowa, Wiley-Blackwel Publishing, USA, page741-772.

Ramsay EC (2014) Rhinoceroses. In West G, Heard D, Caulkett N (eds) Zoo Animal and Wildlife Immobilization and Anesthesia (2nd edn) Ames, Iowa, Wiley-Blackwel Publishing, USA, page635-646.

Redig PT, Duke GE (1976) Intravenously administered ketamine HCl and Diazepam for anesthesia of Raptors. Journal of the American Veterinary Medical Association, 169: 886-889.

Renecker LA, Olsen CD (1986) Antagonism of xylazine HCl with yohimbine HCl and 4-amino pyridine in captive wapiti. Journal of Wild Life Diseases, 22(1): 91-96.

Richard AH (1973) Neuromuscular blocking effect of amino glycoside antibiotics in nonhuman primates. Journal of the American Veterinary Medical Association, 163(6), 613-616.

Richardson KC, Cullen LK (1981) Anesthesia of small kangaroos. Journal of the American Veterinary Medical Association, 179: 1162-1165.

Riebald TW, Kaneps AJ, Schmotzer WB (1986) Reversal of xylazine-induced sedation in llamas, using doxapram or 4-amino pyridine and yohimbine. Journal of the American Veterinary Medical Association, 189(9): 1059-1061.

Richter AG (1977) Ketamine-xylazine immobilization of mule deer. Journal of American Veterinary Medical Association, 171(9): 987.

Ross LG (2001) Restraint, anesthesia and euthanasia. In Wildgoose WH edited, BSAVA Manual of Ornamental Fish, 2nd edn, Gloucester, BSAVA. page75-83.

Ross LG, Ross B (1984) Anesthetic and sedative techniques for fish. Glasgow, Nautical press.

Sabapara RH (1989) Operation immobilization for translocation of a lion. 4(3): 15-16. Zoos' Print, In: A Compendium of publications from Indian Zoos, Vol II, 2000 (Health and Disease Management), Compilers Acharjyo LN, Patnaik SK, Indian Zoo Directors' Association and Central Zoo Authority, page225-226.

Sarma KK, Barthakur T, Thakuria D, Bonal BS, Barua M (1997) Chemical immobilization of Blue Bull (Boselaphus tragocamelus) with ketamine-xylazine mixture and its reversal with yohimbine hydrochloride 12(1): 29. Zoos' Print, In A Compendium of publications from Indian Zoos, Vol II, 2000 (Health and Disease Management), Compilers Acharjyo LN, Patnaik SK, Indian Zoo Directors' Association and Central Zoo Authority, page247.

Sarma KK, Dutta B, Bonal BS (1997) Combination of chemical and physical restraint to treat an injured Rhinoceroses in the Kazi Ranga National Park. 12(6): 36-37. Zoos' Print, In A Compendium of publications from Indian Zoos, Vol II, 2000 (Health and Disease Management), Compilers Acharjyo LN, Patnaik SK, Indian Zoo Directors' Association and Central Zoo Authority, page248-249.

Scarabelli S, Di Girolamo N (2022) Chelonian Sedation and Anesthesia. Veterinary Clinics of North American Exotic Animal Practice, 25(1): 49-72. [doi:10.1016/j.cvex.2021.08.009]

Schmidt MJ (1983) Antagonism of xylazine sedation by yohimbine and 4-amino pyridine in an adult Asian elephant (Elephas maximus). Journal of Zoo Animal Medicine, 14: 94-97.

Schoettger RA, Julin AM (1968) Efficacy of quinaldine as an anesthetic for seven species of fish. Report number 22, US Fisheries and Wildlife services, page9.

Schumacher J (2008) Side effects of etorphine and carfentanil in nondomestic Hoof stock. In Fowler ME, Miller RE (eds) Zoo and Wild Animal Medicine: Current Therapy (Volume 6) St. Louis, Missouri, USA, Saunders Elsevier, page455-461.

Seal US, Schmiti SM, Peterson RO (1985) Carfentanil and xylazine for immobilization of moose (Alcesalces) on Isle Royale. Journal of Wild Life Diseases, 21: 48-51.

Sheelings TF, Holz P, Haynes L and others (2010) A preliminary study of the chemical restraint of selected squamate reptiles with alfaxalone. Proceedings of the Annual Conference of the Association of Reptilian and Amphibian Veterinarians, South Padre Island, Texas, page114-115.

Sontakke SD, Umapathy G, Shivaji S (2009) Yohimbine antagonizes the anesthetic effects of ketamine-xylazine in captive Indian wild felids. Veterinary Anesthesia and Analgesia, 36(1): 34-41.

Spigel RA, Lane TJ, Larsen RE, Cardeilhac PT(1984). Diazepam and succinylcholine chloride for restraint of the American alligator (Alligator mississippiensis). Journal of the American Veterinary Medical Association, 185(11): 1335-1336.

Standard Operating Procedure for the Study of Bats in the Field https://www.nps.gov/subjects/bats/index.htm (This guide provides a good overview of restraint methods for field studies)

Summerfelt RC, Smith LS (1990) Anesthesia, surgery and related techniques. In Methods for Fish Biology, Schreck CB, Moyle PB, Bethesda MD (eds), American Fisheries Society, page213-272.

Thayer CB, Lowe S, Rubright WC (1972) Clinical evaluation of a combination of droperidol and fentanyl as an anesthetic for the rat and hamster. Journal of the American Veterinary Medical Association, 161(6): 665-668.

Thienpont D, Niemegeers CJE (1965) Propoxate (R7467): A new potent agent in cold blooded vertebrates. Nature, 25: 1018-1019.

Thomas DW, William A, Donald DS (1981) Fentanyl and azaperone produced neurolept analgesia in the sea otter (Enhydra lutris). Journal of Wild Life Diseases, 17(3): 337-341.

Thurmon JC, Short CE (2007) History and overview of veterinary anesthesia. Tranquilli WJ, Thurmon JC, Kurt A, Lumb & Jones Veterinary Anesthesia and Analgesia (4th edn) Blackwell Publishing Professional, page3-6.

Viviana Q, Fernandes-Santos RC (2014) Tapir Veterinary Manual (2nd edn) IUCN/SSC, Tapir Specialist Group (TSG), www.tapirs.org

Wallach JD and Boever (1983) Diseases of exotic animals. Medical and surgical management, WB Saunders and Company, Philadelphia, USA, page1159.

Wallach JD (1966) Immobilization and translocation of the White square lipped rhinoceros. Journal of the American Veterinary Medical Association, 149(7): 871-874.

Walner BM, Hatch RC, Booth NH, Kitzmzn JV, Clark JD, Brown J (1982) Complete immobility produced in dogs by xylazine atropine antagonism by 4-amino pyridine and yohimbine. American Journal of Veterinary Research, 43: 2259-2265.

Walzer C, Greisman H (2015) Update on remote delivery and restraint equipment. In Miller RE, Fowler ME (eds) Zoo and Wild Animal Medicine (Volume 8) St Louis, Missouri Elsevier Saunders, page730-732.

Wenger S (2012) Anesthesia and analgesia in rabbits and rodents. Journal of Exotic Pet Medicine, 21: 7-16.

Williams BH, Kathy A, Huntington B, Miller M (2018) Mustelids. In Terio KA, McAloose D, Leger JS (eds) Pathology of Wildlife and Zoo Animals, Academic Press, ISBN 012809219X, 9780128092194, 1136 pages.

Williams TD, Kocher FH (1978) Comparison of anesthetic agents in the sea otter. Journal of the American Veterinary Medical Association, 173(9): 1127-1130.

Wolf A (1970) Immobilization of captive and free-ranging white tailed Deer (Odocoileus virginianus) with etorphine HCl. Journal of the American Veterinary Medical Association, 175(5): 636-640.

Wolf A, Swart JH (1970) Etorphine HCl: Anesthetic for surgery on an elk (Cervus Canadensis Canadensis). Journal of the American Veterinary Medical Association, 157(5): 641-642.

Wolfe B (2015) Bovidae (except sheep and goats) and antilocaprid. In Miller RE, Fowler ME (eds) Zoo and Wild Animal Medicine (Volume 8) St Louis, Missouri, Elsevier Saunders, page626-644.

Appendix

Sl. No	Generic name	Trade name	Manufacturer supplier
1.	Acepromazine maleate	**Acetylpromazine** **Acepromazine** (25mg/ml) (50ml Vial)	*Boots Pure Drugs Co, Station Street, Nottingham, England *NexGen ANIMAL HEALTH, 953Hilltop Dr, Weatherford, TX70686
2.	Acepromazine maleate+Etorphine HCl	**Immobilion** Large animal (LA) [Etorpine HCl 2.45mg/ml and Acepromazine maleate 10mg/ml]	*American Cyanamid Co., PO Box 400 Princeton, New Jersey 08540, USA *Reckitt and Colman, Dansom Lane Hull, HU8 7DS, England *Novartis Animal Health UK Ltd.
3.	Alphaxolone +Alphadolone acetate	**Saffan** **Althesin**	*Glaxo Laboratories Ltd., Green Ford, Middlesex, UB60HE, England
4.	Azaperone	**Serenic / Stresnil** **Azaperone** (40mg/ml) (50ml Vial)	*Janssen Pharmaceutica, Research Laboratories, Beerse, Belgium *Livealth Biopharma Private Limited, Kopar Khairane, Navi Mumbai, Thane, India
5.	Chlordiazepoxie	**Librium**	*Roche Laboratories, Hoffman-La Roche Inc., Nutley, NJ 97110, USA

6.	Chlorpromazine HCl	**Largactil/ Megaphen/ Thorazine** (25mg/ml) **Chlorpromazine** (2.5%, 2ml) **Megatil** (25mg/50mg/100mg/ 200mg)10ml vial	*Roche Laboratories, Hoffman-La Roche Inc.,Nutley, NJ 97110, USA *Pitman-Moore Co., PO Box 344, Washington Crossing, NJ 08560, USA *Piramal Pharma Limited, LBS Marg, Kurla West, Mumbai, India *Intas Pharmaceuticals Limited, Corporate House, Near sola Bridge, Thaltej, Ahmedabad, Gujrat, India
7.	Cyprenorphine HCl	**M-285**	*Reckitt and Colman, Dansom Lane Hull, HU8 7DS,England
8.	Diazepam	**Valium** **Calmpose** (10mg/2ml)	*Roche Laboratories, Hoffman-La Roche Inc., Nutley, NJ 97110, USA *Roche Products, Welwyn Garden city Herts, England *Ranbaxy Laboratories Ltd., Plot No 90, Sector 32, Gurugram, Haryana, India
9.	Diethyl thiambutene HCl	**Themalon**	*Burroughs Wellcome & Co., 183-193 Euston Road, London, NWI, England
10.	Dihydromorphine HCl	**Dilaudid** (1mg/ml, 2mg/ml, 4mg/ ml, 10mg/ml)	*Knoll AG67, Ludwighshafenam Rhein, Post Fach 210805, Germany *Knoll Pharmaceutical Co, Arrange, New Jersey 07000, USA
11.	Diprenorphine HCl	**Revivon/ M 5050**	*American Cyanamid Co., PO Box 400 Princeton, New Jersey 08540, USA *Reckitt and Colman, Danson Lane, Hull, HU8 7DS, England
12.	Droperidol	**Droperidol/** Haloperidol / Inapsin (5mg/2ml)	*Janssen Pharmaceutica, Research Laboratories, Beerse, Belgium *Pitman-Moore Co., PO Box 344, Washington Crossing, NJ 08560, USA
13.	Droperidol-fentanyl	Innovar-vet thalamanol	*Janssen Pharmaceutica, Research Laboratories, Beerse, Belgium *Pitman-Moore Co., PO Box 344, Washington Crossing, NJ 08560, USA

14.	Enflurane	**Ethrane** (125ml/250ml) **Enflurane USP** 99.9% (250 ml) **Isoflurane USP** 99.9% (250 ml) **Forane** (Isoflurane) (250 ml)	*Abbott Laboratories, North Chicago, Illinois 60064, USA *Abbot Laboratories, Queen Borough, Kent, England *Piramal Pharma Limited, LBS Marg, Kurla West, Mumbai, India *Baxter Healthcare Corporation Deerfield, IL 60015, USA
15.	Etorphine HCl	**M-99** (10mg/10ml) 30ml vial	*American Cyanamid Co., PO Box 400, Princeton, New Jersey 08540, USA *Reckitt and Colman, Dansom Lane Hull, HU8 7DS, England
16.	Etorphine HCl + Acepromazine maleate	**Immobilan LA**	*American Cyanamid Co., PO Box 400, Princeton, New Jersey 08540, USA *Reckitt and Colman, Dansom Lane Hull, HU8 7DS, England
17.	Etorphine HCl + Ethotrimeprazine	**Immobilan SA**	*American Cyanamid Co., PO Box 400, Princeton, New Jersey 08540, USA *Reckitt and Colman, Dansom Lane Hull, HU8 7DS, England
18.	Fentanyl citrate	**Sublimaze** (0.05mg/ml) 5ml/10ml/20ml vial **Fentanyl** (50 mcg/ml) 2ml Ampoule	*Janssen Pharmaceutica, Research Laboratories, Beerse, Belgium *Piramal Pharma Limited, LBS Marg, Kurla West, Mumbai, India *Hikma Pharmaceuticals,USA Inc., Cherry hill, New Jersy, 08003
19.	Fentanyl citrate + Fluanisone	**Hypnorm**	*Janssen Pharmaceutica, Research Laboratories, Beerse, Belgium
20.	Gallamine triethiodide	**Flaxedil** (100mg/ ml) 2ml Ampoule	*May and Baker Ltd., Dagenham Essex, England *Pitman-Moore Co., PO Box 344, Washington Crossing, NJ 08560, USA

21.	Halothane	**Fluothane/ Halothane** 250 ml Bottle **Halothane B.P.** 250 ml Bottle **Halothane B.P.** 250 ml Bottle	*ICI American Inc., Atlas International Division, Wilmington, Delaware 19899, USA *ICI Pharmaceuticals, Alderley House, Alderley Park, Mackles Field, Cheshire, SK1047F, England *Raman & Well Pvt., Ltd., Marine Road, Mumbai, Maharashtra, India *Taj Generics Pharmaceuticals Limited, Oshiwara Industrial Centre, Goregaon West Mumbai, India
22.	Hexobarbitone Sodium	**Evipan** **Citopan/Tobinal**	*Bayer Products Limited, Africa house, Kingsway, London UK *Winthrop Laboratories, 1450 Broadway, New York 10012 USA
23.	Ketamine HCl	**Vetalar/ Ketalar** (50mg/ml), 2ml/10ml (10mg/ml) 20ml vial **Ketamin** (10mg/ml) 10ml/20 ml (50mg/ml), 2ml/10ml **Aneket** 100mg in 2ml ampule 250 mg in 5 ml vial 500mg/1000mg in 10 ml vial **Anesket** (1000mg/10ml) **Ketapil** (50mg/ml) 10ml vial	*Parke Davis & Co., Josep Campau Avenue at-the River Detroit, Michigan 48232, USA *Parke Davis & Co. Ltd., USK Road, Pontypool, Gwent, NP 8YH, UK *Themis Medicare Ltd., Udyog Nagar S.V.Road, Goregaon (W), Mumbai, Maharashtra, India *Neon Laboratories Limited, Gundavali, Andheri East, Mumbai, Maharashtra, India *PiSA Laboratories, Urbana Zacatecas, Mexicali B.C., Mexico *Psychotropics India Limited, Salempur, Haridwar, Uttarakhand, India
24.	Levallorphan tartarate	**Lorfan**	*Roche Laboratories, Hoffman-La Roche Inc., Nutley, NJ 97110, USA *Roche Products,Welwyn Garden City, Herts, England

25.	Lignocaine HCl	**Xylocaine/ Xylotox** (2%) **Lox** (2%) 30ml vial **Lignocaine** (2%) 30ml vial **Lignox** (2%, 4%) 30ml vial **Xylocard** (2%) 50ml vial	*Astra Pharmaceutical Products, Worcester Massachusetts 01600, USA *Willows Francis, Westhoughton, Epsom Surrey, England *Neon Laboratories Limited, Gundavali, Andheri East, Mumbai, Maharashtra, India *Cadila Pharmaceutical Limited, Sarkhej-Dhokla Road, Bhat, Ahmedabad, Gujrat, India *Indoco Remedies Limited, Indoco House, 166 CST Road, Santacruz (E), Mumbai, India *Astra Zeneca Pharma India Limited (AZPIL), Manyata Tech Park, Banglore, India
26.	Mephensin	**Myanesin/ Tolserol**	E.R.Squibb and Sons, 74-5, 5th Avenue, New York 10022, USA
27.	Methoxyflurane	**Metofane/ Penthrane** 125 ml	*Abbott Laboratories, North Chicago Illinois, 60064 USA *Abbott Laboratories, Queen Borough, Kent, England *Abbott India Limited, 3 Corporate Park, Chembur, Mumbai, Maharashtra, India
28.	Metomidate (methoxymol)	**Hypnodil/ R7315** **Hypnodate** (2mg/ml) 2ml	*Janssen Pharmaceutica, Research Laboratories, Beerse, Belgium *Neon Laboratories Limited, Gundavali, Andheri East, Mumbai, Maharashtra, India
29.	Nalorphine HCl	**Lethidrone** (0.4mg/ml) 1ml/10ml **Naloxone Hydrochloride** (0.4mg/ml) 1ml/10ml	*Burroughs Wellcome & Co., 183-193 Eusfon Road, London, NWI, England *Taj Generics Pharmaceuticals Limited, Oshiwara Industrial Centre, Goregaon West Mumbai, India *Hospira health care India Private limited, GN Chetty road, Chennai, India

30.	Naloxone HCl	**Naloxone** (0.4mg/ml)	*Endo Laboratories, Australia Private Ltd., 448, Pacific Highway, Astron, New South Wales, Australia *Endo Laboratories Inc., 1000 Stewart Avenue, Garden City, New York 11530, USA
31.	Pentobarbitone	**Somnotol/ Euthanyl/ Socumb/ Pentobarbital Sodium/Pentobarbital** (65mg/ml)	*May and Baker Ltd., Dagenham, Essex England *Pitman-Moore Co., PO Box 344, Washington Crossing, NJ 08560, USA
32.	Pentobarbitone sodium+Chloral hydrate+Magnesium sulphate	**Equithesin** Pentobarbitone sodium 8.86mg+ Chloral hydrate 42.5mg+Magnesium sulphate 21.2mg/ml	Jensen-Salsbery Labs In, 520, West 21st Street, Kansas City, Missouri, 64141, USA
33.	Phencyclidine HCl	**Sernylan/ Sernyl**	*Bio-Ceutic Laboratories Inc., PO Box 999 St. Joseph, Missouri 64-502,USA
34.	Procaine HCl	**Novocaine** (1% or 2%) **Planocaine** (1% or 2%) **Procainamide Hydrochloride** (100mg/ml) 10ml vial	*Sterling Drugs International, Sterling House 14, Hewlett Stree, Jondon, EC2 A3NJ, England *Winthrop Laboratories, 1450 Broadway, New York 10012, USA *Hospira Inc., Lake Forest, IL 60045, USA
35.	Promazine HCl	**Primazine/Prozine-50 / Sparine** 50mg/ml	*Fisons Ltd. Pharmaceutical Division, Loughborough, Leicestershire, LE129 TE UK *Wyeth Laboratories, PO BOX 8299, Philadelphia, Pennsylvania 19101, USA
36.	Propionyl promazine HCl	**Tranvet**	Diamond Laboratories Inc., 2538 South East 43rd Street,Des Maines, IOWA 50304, USA
37.	Reserpine	**Serpasil** 0.1 /0.25mg **Serpalan** 0.1 /0.25mg	*Ciba-Geigy Ltd., Basle, Switzerland *Ciba Pharmaceutical Co., Summit NJ07901 USA

38.	Succinyl Choline	**Midarine** (50mg/ml) 2ml **Scoline** (50mg/ml) 2ml **Suxomine** (50mg/ml) 10ml	*Burroughs Welcome Pharmacia Private Limited, now GlaxoSmithKline Pharmaceuticals, 252, Dr Annie Besant Road, Worli, Mumbai, India *GlaxoSmithKline Pharmaceuticals, 252, Dr Annie Besant Road, Worli, Mumbai, India *VHB Life Sciences limited, Kandivali West, Mumbai, Maharashtra, India
39.	Tetracaine HCl	**Pentocaine** **Tetracaine Hydrochloride** Opthalmic Solution 0.5% in 10 ml	*Haver-Lockhart Labs, Kansas City, Missouri 64100, USA *Amici Pharmaceuticals, Melvilli, New York, USA
40.	Thiamylal sodium	**Surital/Anestatal** 1gm/5gm	*Parke Davis & Co. Ltd., USK Road, Pontypool, Gwent, NP 8YH, UK
41.	Thiopentone sodium	**Intraval Sodium** (500mg/ 1gm) **Thiosol** (500mg/ 1gm) **Pentothal** (500mg/ 1gm) **Intraval Sodium** (500mg/ 1gm)	*May and Baker Ltd., Dagenham, Essex England *Neon Laboratories Limited, Gundavali, Andheri East, Mumbai, Maharashtra, India *Pharmacia & Upjohn Pfizer Inc., 235 East, 42nd Street, NewYork *Piramal Pharma Limited, LBS Marg, Kurla West, Mumbai, India
42.	Tiletamine HCl	**CI-634**	*Parke Davis & Co., Josep Campau Avenue at-the River, Detroit, Michigan 48232, USA

43.	Tiletamine HCl + Zolazepam	**CI-744** **Zoletil 50** Tiletamine HCl 125mg + Zolazepam125 mg) **Zoletil 100** (Tiletamine HCl 250mg+Zolazepam 250 mg) **Telazol** 100mg/ml total (Tiletamine HCl 50mg+Zolazepam 50 mg)	*Parke Davis & Co., Josep Campau Avenue at-the River, Detroit, Michigan 48232, USA *Virbac Animal Health India Pvt. Ltd., Western Express Highway, Borivali East, Mumbai, Maharashtra, India *Virbac Animal Health India Pvt. Ltd., Western Express Highway, Borivali East, Mumbai, Maharashtra, India *Zoetis India Limited, Kalina, Santacruz East Mumbai, Maharashtra, India
44.	Tricaine methane sulphonate	**MS-222/Finquel/ Tranquil/Tricaine-S/ Syncaine** (1000mg/gm)	*Sandoz Ltd., Sandoz House, 98, Thecentre Felthum, Middle Sex Tw134EP, England *Sandoz Pharmaceuticals, Hanover, New Jersey 07936, USA *Syndel, 1441 W Smith Road, Ferndale, Washington
45.	Trichlorethylene	**Trilene** 250 ml Bottle	*ICI American Inc., Atlas International Division, Wilmington, Delaware 19899, USA *ICI Pharmaceuticals, Alderley House, Alderley Park, Mackles Field, Cheshire, SK1047F, England
46.	Triflupromazine HCl	**Siquil** (20mg/ml) 5ml vial **Sicvel Vet** (20mg/ml) 5ml vial **Triquil** (20mg/ml) 5ml vial **Siquil**	*Zenex Animal health India Pvt. Ltd., Zydus Corporate Park, Ahmedabad, Gujrat, India *Morvel Laboratories (P.) Ltd., Mahesana, Gujrat, India *Puremed Biotech, Sai Baddi, solan, Himachal Pradesh, India *24 E.R. Squibb and Sons, 74-5, 5th Avenue Newyork, 10022, USA

47.	Xylazine HCl	**Rompun** **Xylocad** (20 mg/ml) 2ml/10ml **XYLO-B** (20 mg/ml) 30ml vial **Xylaxin** (20 mg/ml) 2ml/10ml/30ml vial	Diamond Laboratories Inc., 2538 South East 43rd Street, Des Maines, IOWA 50304, USA *Cadilla Pharmaceuticals Limited, *Brilliant Bio Pharma Private Limited, Anirich Industrial Estate, Bollaram, Telangana, India *Indian Immunologicals Limited, Hyderabad, Telangana, India

Index